A Brief History of Physical Science

A Brief History of
Physical Science

John A. Cramer

Writers Club Press
San Jose New York Lincoln Shanghai

A Brief History of Physical Science

Writers Club Press
an imprint of iUniverse.com, Inc.

For information address:
iUniverse.com, Inc.
5220 S 16th, Ste. 200
Lincoln, NE 68512
www.iuniverse.com

ISBN: 0-595-19754-X

Printed in the United States of America

Contents

Preface

This book has developed over many years of teaching core physical science courses at Oglethorpe University. It is aimed at an audience unfamiliar with the physical sciences in any organized way. Previous experience had taught me the value of teaching the subject from an historical perspective. So this book is a brief history of the subject. I say brief because I have made no attempt to touch on every topic in the physical sciences. I have not even tried to be comprehensive within the topics I have chosen. Instead, I have tried to focus on the primary and crucial developments, the changes of thinking and the essential discoveries, that have molded and shaped the physical sciences into what they are today.

Physics and chemistry textbooks are designed to teach the essential contents of the subjects as concisely and economically as possible. The sciences are by now so detailed and a knowledge of their applications so necessary in the preparation of science majors that the textbooks have no room for showing how the details came to be what they are today. To do that a different pedagogy is required. Studying the historical development of the sciences is far better suited to the task.

It is useful to see this development as a rope of many strands twisted together. Each chapter of the book traces a different strand as it winds through the rope. Most of the strands originate in classical Greek thought and so that is where most of the chapters begin. The first two strands are attitudinal and the other five are content strands. Science does not stand alone but grows only in prepared soil (to mix metaphors). Cultural attitudes have clearly been crucial to the development of science in Western Europe. We will not see science for what it is if we remain unaware of these attitudinal strands.

Most books on the history of science are organized by century or time period. This book traces each topic more or less separately over time. The reader may at points think this organization of material inefficient. The choice is basically pedagogical. It is easier to follow the development of a single topic than of many at once. Think of a rope again. The best, perhaps the only, way to see that the strands at one end match those at the other is to trace each strand individually down the length of the rope. Many fine histories of science are organized by period but in teaching clarity must be prized. I am convinced, from years of teaching the subject, that the organization of this book makes the subject easier to follow for most readers.

Introduction

Most of us have no idea why we believe matter is made of atoms and the Earth goes around the Sun. Fundamental as these "facts" are, few of us can cite evidence to show they are true. We just know that experts know why and there is nothing more we can say about it.

While it would be nice if we were all better informed, we all have to depend on experts in one area or another. The scope of human knowledge is just too broad for any of us to be expert at everything. In fact, depending on experts is a necessary factor in the progress that science has made over the last two millennia. Trusting the work of others allows us to stop "reinventing the wheel."

This is a book about how we have reached much of our current understanding of the physical universe. The subject is a rope of many strands, a story with many subplots. Each chapter traces out a strand of the rope. But in all of them, trust has been basic. Trust is part of the fabric of the strands. Science is open; the data and work of any scientist is available and open to the examination of all other scientists. Trust might not seem all that necessary. But in actual practice, scientific reports are examined closely by a very small number of people. The vast body of scientists takes a scientific report at face value, on trust. And even the few examiners assume a basic honesty in the report. Fraud does occur and it is almost always caught when someone involved in or close to the original work "blows the whistle" on it.

Thus, what Jacob Bronowski has called "the habit of truth" has been a prerequisite to the development of science in the West. This sense of an objective truth independent of oneself or one's culture or society is unusual, possibly an exclusive feature of recent Western Europe. It is absent from almost all other cultures and times. It may well be disappearing now in the West. If it is lost, the repercussions cannot be fully foreseen but it seems most likely that scientific progress will not persist.

There are other prerequisites for scientific progress. The habit of truth is closely bound with two other attitudinal strands of modern science. People had to have the courage to believe both that the universe has a mathematically precise order and that experiments can lead to valid knowledge about the world. Like the habit of truth, these attitudes had to be learned and other attitudes discarded. The three attitudes support each other. We will see, for example, Johannes Kepler throwing away a precise mathematical theory of the motion of Mars when it does not fit Tycho Brahe's data on Mars *to within the precision of the data*. In that decision, all three attitudes combine. For that reason, Arthur Koestler called Kepler's decision a watershed of science, a step that set the direction of all later steps.

Other strands were the growing understanding of the natural order in specific content areas: motion on Earth and in the heavens, matter, energy, light, and electricity and magnetism. In all these areas, centuries of thinking and culling, experimenting and theorizing, have led us to the current status. Sometimes the options have been reasonably clear from the outset. Over time, accumulating evidence selects one from the two early theories of matter. For other strands, nothing is clear at the outset. Electricity and magnetism were far too puzzling for early thinkers either to make much of them or to determine what might explain them. Energy was not even foreseen by the great Isaac Newton, let alone the Greeks.

In these content areas of modern science, the importance of the attitudinal strands is evident as their stories unfold. Once the proper attitudes are in place, the content areas blossom and expand, yielding their secrets to experiment and mathematical analysis.

Other strands also appear. Scientists are people of their time and science is a human endeavor. No less than others, scientists are influenced by their times. Cultural factors of all sorts, philosophical, religious, social, political, and economic, have always influenced the development of science. We will see many of these factors and their effects in the course of tracing out the story of how modern science got to where it is today.

Chapter I

Belief in a Quantifiable Order of the Universe

The Classical Period

The beginnings of physical science can be assigned somewhat arbitrarily to the Classical period of Greece. Especially in Athens, a speculative but rational way of thinking about many topics developed. Politics and truth, the good and the beautiful, the world and things in it were all discussed in a new way reminiscent of modern physical science. For that reason, we regard classical Greece as the birthplace of physical science.

Why this new attitude arose is a complex question. One suggested answer is that usually only upper class men with leisure time engaged in this speculative activity, so the birth of modern science and philosophy is sometimes ascribed to the rise of the leisure class. If that is correct, it would be only one of many times when the development of science was affected by shifting cultural patterns. Arguably, the crucial contribution of classical

Greece to modern science has been the belief that there is indeed a rational order to the universe. Some ancient Greeks taught that rational people can comprehend this rational order. They began the search for that order.

Of course, other peoples of that time, even before that time, had believed that the universe is ordered. The monotheistic Jews saw a universe ordered by fiat of an all-wise, all-knowing God. Even their polytheistic neighbors saw a kind of order in the universe, although it was a more confused and tumultuous order paralleling the overlapping responsibilities and mixed motives of their pantheon of gods. In contrast, these Greeks thought the cosmos was mathematical ordered, an order based on organizing principles.

Mathematics and the application of mathematics did not begin with the Greeks. Numbers and arithmetic arose out of practical counting needs in most cultures. Names for the integers are part of every language and examples of simple arithmetic appear in many of the written records available to archeologists. Geometry arose out of the practical needs of land measurement in Egypt, where the Nile annually altered farm plots. It also arose in Babylon out of the need of accuracy in measuring positions of stars and planets for astrological purposes. But it was the Greeks who first systematized and organized mathematics along the abstract and general lines that characterize modern mathematics.

Thales (624-548 BC) of the Greek city Miletus in Ionia (now the Turkish coast) learned geometry in Egypt. Returning to Miletus, he began to teach. In his hands, geometry became deductive, proofs replaced demonstration, and mathematics became abstract. The deductive proofs of Thales lent a certainty to his mathematical conclusions which later peoples, wondering about the certainty of human knowledge, found very attractive. Thales saw the order of the cosmos in terms of the single element, water, of which all matter was supposedly composed. Thus, he became the first person to attempt to organize an understanding of the universe around a few naturalistic principles.

Another Ionian, Pythagoras (c.575-495 BC), made mathematics the basis of a metaphysical view where numbers were the foundation of all reality, the appearances of the world having beneath them the true reality of numbers. He founded a school at Kroton, in southern Italy, and gathered about him a quasi-religious society of disciples devoted to studying mathematics and to showing how mathematics can be used to explain and understand the world.

These Pythagoreans held that an object cannot be infinitely divided but can only be divided into a finite number of pieces. An object can be assigned a number equal to the number of its pieces (or, perhaps, the number of pieces in its outline). They discovered that the harmonics of a string, of musical notes, are integer multiples of the fundamental tone. Harmony based on integer ratios became a major theme of their thinking. For them, harmony applied to music, architecture, the cosmos, and even medicine and health.

In their mystical system, numbers constituted the essence of things. 1 was not a number, but the basis of all numbers. It represented reason. The number 2 represented women and 3 represented men, so 5 represented marriage. Because it was the sum of equals (2 and 2), 4 represented justice. The Pythagoreans preferred to represent numbers with triangles of dots. A triangle of 1, 2, 3, and 4 dots in four rows represented 10, which had a "sacred fourfoldness" because it had four rows ending in four dots. Thus, 10 became the sacred number (*tetraktys*).

Perfection was another important theme of Pythagorean thought. The circle was the most perfect form because it is the same, unchanging, regardless of what side it is viewed from. To their way of thinking "unchanging", "constant", and "perfect" were interchangeable terms because "something that is perfect cannot be improved so it must remain constant"[i]. They regarded the underlying numbers and geometric forms as perfect, independent of the imperfect appearances in the world.

The Pythagorean Theorem was the major achievement of Pythagoras but, ironically, it also caused the downfall of his metaphysical program because it led to the discovery of irrational numbers. Since the Pythagoreans conceived the universe in terms of counting and integer numbers, a scheme with no room for irrational numbers, the discovery brought the Pythagorean program to a halt. The theorem is that, in right triangles, the square of the hypotenuse equals the sum of the squares of the two other sides. In equation form that becomes $c^2 = a^2 + b^2$ where a, b, and c are defined in Figure 1.

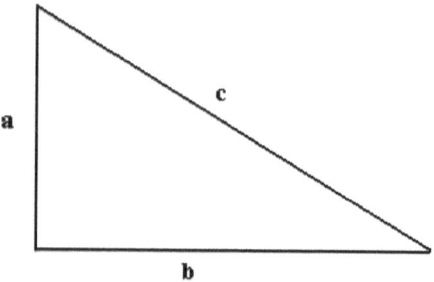

Figure 1. The Pythagorean Theorem

Irrational numbers are numbers that cannot be represented by a ratio of whole numbers (a proper fraction). They arise easily in the geometry of right triangles. For example, a triangle with each side equal to 1 has an hypotenuse of $\sqrt{2}$, an irrational number. Legend has it that the discovery of irrational numbers was made by one of the Pythagorean brotherhood in the course of a journey with other Pythagoreans in a boat. The discovery so distressed his fellows that they threw him overboard!

Pythagoras and his students were the first to believe that there is order in the universe accessible to mathematical analysis. The order they expected was based on the counting numbers (the natural numbers or positive integers) and ratios of the counting numbers (rational numbers).

When their program foundered on the discovery of irrational numbers, the theory of numbers and geometry were forced to develop separately from that time until the seventeenth century when René Descartes recombined them in analytic geometry. With Plato and then Aristotle, confidence in mathematical order declined.

Plato (428-347 BC) began the attempt to make philosophy deductive in a manner analogous to the mathematics of Thales and Pythagoras. He extended the Pythagorean idea of ideal forms underlying reality. To him the forms (Platonic forms or Platonic ideals) were the true and trustworthy reality of which the world of appearances was an untrustworthy reflection. Irrational numbers were not a problem because Plato viewed the order of the universe in geometrical terms rather than the arithmetical terms of the Pythagoreans. Irrational numbers are generated by geometry and seem natural in that context. Plato's tentative view of the nature of the universe is set out in his book the *Timaeus*. He called the view only a "likely story" in order to disclaim certainty for the scheme but he was serious enough about it to record it. In the *Timaeus* we see there is order in nature which reason can comprehend. The world of appearances is a likeness or image (eikon) of the underlying perfect forms. Thus, it has a degree of perfection and order. The world of appearances is like a work of art fashioned by an artisan (the "Demiurge") out of the chaos of materials available so it has chaotic tendencies that make it imperfect. The world of forms is perfect and hence will reward rational study better than will the world of appearances. That is the point of the Allegory of the Cave from *The Republic* where the shadows on the wall (the world of appearances) reflect as well as distort the underlying reality (the actors behind the observers).

Plato applied mathematics to the forms. He believed everything on Earth is made of the four elements fire, air, water and earth and each element was made of "atoms" of a specific geometric form. The "atoms" themselves were constructed of right triangles that functioned rather like the atomic particles of modern atomic theory. His colleague Theatatus having recently proven that only five regular solids are possible, Plato appropriated the solids. Since the sides of four of them could be bisected to make triangles, these four solids, he thought, must represent the four elements. The last solid, the dodecahedron must then represent the cosmos (heavens).

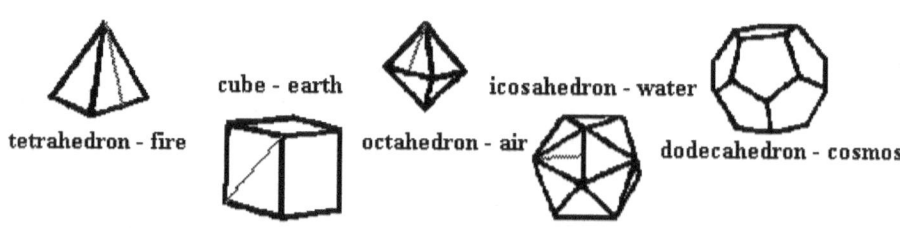

Figure 2. The five Platonic Solids. Note the faces of the first four can be divided into two equal right triangles.

Mathematics was expected to be especially useful in describing the heavens because they were more perfect than things on Earth. He began the pattern of treating the motions of heavenly bodies as combinations of circles and set the famous "Plato's Problem" for his students: find a set of circles which will replicate the observed celestial motions.

Plato thought studying the forms would elevate the mind, an attitude which eventually led to a neglect of, even an aversion for, the material world. This neglect was amplified by the rising perception among the leisure class that work with the hands was a slave's job. One of the hazards of a slave society is this sort of devaluation of craftsmanship and "honest" work. After Plato's death, his Academy split into those who wanted to study the Forms (nicknamed the "gods") and those wanting to study the material world (the "giants").

Aristotle (384-322 BC) became the greatest of the giants. Logical problems with the identification of form with number led him to conclude that "mathematics could only treat of primary movement, that is, movement in its simplest, eternal form-rotation of a circle. Therefore, mathematics could be used to describe heavenly rotation but not the variable rectilinear movements of earthly physics."[ii] For Aristotle, order was based on purpose and causality rather than on mathematics. Both for Plato and Aristotle, logic was critical to an understanding of the world. Aristotle went so far as to call logic the tool (organon) of knowledge (science). He developed a more complex version of Plato's cosmology in which the universe was divided into the celestial region and the earthly region. The lunar orbit was the dividing line between the two regions and the Moon itself was a part of the celestial region.

The universe was comprehensible to human reason but order on Earth could only be described qualitatively; mathematics was useless. Aristotle's lack of confidence in the order of the earthly sphere forced him into classifying material phenomena, a qualitative procedure, and turned him toward biology rather than the physical sciences.

The standards of rigorous (deductive) reasoning in mathematics were permanently set by Euclid (330-275 BC) in his *Elements of Geometry*. A student of Euclid's, Appolonius (262-200 BC), wrote *Conic Sections* and also invented the eccentric as a device in descriptive astronomy. We will later see it in use in the Ptolemaic system.

The greatest mathematician of antiquity, Archimedes (287-212 BC), was also a great physicist and inventor known for theorems on levers, the discovery of the basic principles of hydrostatics, inventing (or improving) the Archimedean screw for raising water, inventing compound pulleys and, as a young man, building a water powered planetarium. Combining mathematical analysis and experimental observation-based technology in the work of one person is characteristic of modern physical science. Perhaps Archimedes should be called the only ancient, modern scientist.

In mathematics, Archimedes invented a way of writing and calculating with large numbers that he used in his essay, "The Sand Reckoner", to calculate the maximum number of grains of sand the universe could contain. He found the number of grains of sand for a geocentric universe was

10^{51} and 10^{63} for a heliocentric universe. He also used the method of exhaustion (evidently originated by the Pythagoreans) for calculating areas. In particular, he proved the area of a circle is πr^2. His procedure was to divide the circle into many, equal wedges and to approximate each wedge as a triangle whose area he could calculate. Increasing the number of wedges makes the approximation more accurate, and making the number infinite exhausts the area of the circle. His method was useful in the development of the calculus by Leibniz and Newton two millennia later.

The classical period ended around 200 BC, though the classical tradition lingered for a while in Alexandria. The political decline of Athens and the fragmentation of the Alexandrian Empire led to scattering of scholars, who were without journals or the printing press for communication. Furthermore, confidence in the program of quantification was shallowly rooted and seems to have waned in the face of the many obstacles and unsolved problems. The classical Greeks saw how geometry applied to heavenly motion but not even Archimedes was able to imagine applying it to motions on Earth.

Two notable scholars who continued the classical tradition were Eratosthenes and, much later (c.150 AD), Claudius Ptolemy who produced a complete geometric scheme of the cosmos. About 250 BC, Eratosthenes, the librarian of the Alexandrian Museum, calculated the circumference of the Earth. His method (see Figure 3), based on the erroneous information that the Sun was directly overhead at noon in Syene (modern Aswan) at the summer solstice, was to measure the position of the Sun at that time in Alexandria. From this angle and the known distance from Alexandria to Syene he calculated the circumference of the Earth at about 250,000 *stadia*. The size of the *stadium* changed over time so his figure does not convert unambiguously to modern units. And he made it 252,000 to divide easily by 60, the base of his number system. However, his result was no more than 16% (and perhaps only 2%) too big! Later reductions in the size of the *stadium* and Ptolemy's passing on of Posidonius' figure of 180,000 *stadia* in his *Geography* contributed, in the time of Columbus, to a badly under-estimated distance westward to China. Thus, Columbus was encouraged to attempt (and Ferdinand and Isabella were persuaded to finance) the trip which led to the discovery of the new world.

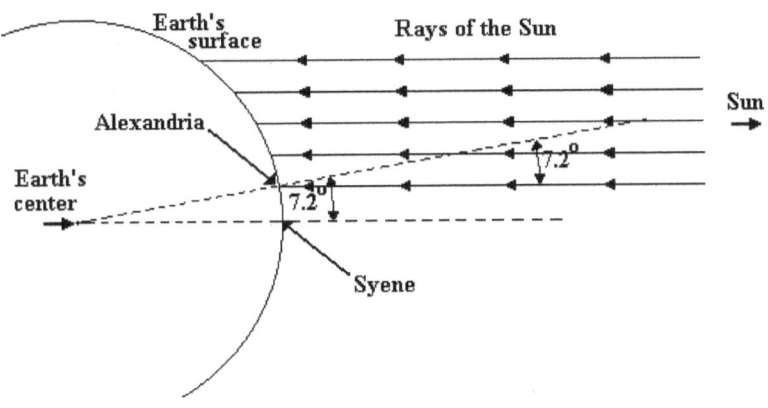

Figure 3. Eratosthenes' Method for Measuring the Earth

The Interlude

After the Classical period, stagnation began in the West for a number of reasons. The decline of Greek power and the subsequent breaking of the chain of teacher to student transfer of knowledge and ideas and the rise of the Roman Empire with its pragmatic outlook all opposed the continuance of contemplative and abstract thinking. The Romans looked for moral lessons or directly useful applications from knowledge. They had no time for the speculative efforts that had fueled Greek scientific thought.

Also, the rise of Christianity, where the focus of abstract thought was theological and otherworldly, diverted intellectual energy away from science and mathematics. There were two branches of opinion within Christianity. The first branch, including Clement and Origen of Alexandria in the second century and St. Augustine of Hippo in the fourth and fifth centuries, admired and used Greek thought primarily for theological purposes and only secondarily for anything like a scientific purpose. The second branch entirely rejected Greek philosophy and science as unnecessary or even contaminating. The late second century church father Tertullian is often quoted in this respect "What has Jerusalem to do with Athens, the Church with the Academy, the Christian with the heretic? ... Let us have faith and wish for nothing more."[iii] The irony is that Tertullian, the son of centurion, was well educated in Greek philosophy and Roman law.

Finally, the barbarian invasions of Europe led to an economic and social collapse that required hundreds of years to overcome. People living at subsistence levels have little time for abstract thought and contemplation. The history of ideas is more one of false starts and dashed hopes than of uninterrupted progress. To expect the last several hundred years of scientific advance to characterize the entire history of science is unrealistic.

In the east, science was maintained and even advanced. The Eastern Roman Empire was not invaded by barbarians so social structures continued the traditions and education of early Greece although the Emperor Justinian closed the Platonic Academy in 529 AD because of its paganism. At that point, most of the scholars fled to Persia where royal support long maintained Jewish and Christian Syriac speaking scholars.

Islam rose in the early seventh century. By the eighth century a vast Islamic Empire stretched from Spain and North Africa to the borders of India. The empire absorbed Greek tradition as well as other traditions. In particular, the Arabs adopted Hindu numbers including the zero but they rejected the Hindu use of negative numbers. They improved the astrolabe (Figure 4.), invented by Hipparchus a millennium earlier, and applied it to guiding camel caravans traveling at night. They made important advances in algebra especially in the use and solution of quadratic equations (al-Khwarizmi) and cubic equations (Omar Khayyam). Our word "algebra" is derived from the Arabic "al-jabr" meaning "the restoration" (of an unknown number).

They also improved the Ptolemaic system (preserving Ptolemy's writings for us as they did so). In fact, Ptolemy's main work, *The Mathematical Composition*, is now known by its Arabic name *Almagest* ("the greatest"). On the whole, however, "their purpose was not to construct a better physical science, but to demonstrate how the science they possessed could lead to the ultimate goal of all knowledge. As it did for the Christian, the transcendental world actually determined for them the goals of this-worldly science." iv

Few improvements in mathematics were made in the non-Islamic world until the Crusades. Some exceptions were Diophantus of Alexandria (320 AD?) who used letters for unknowns and the symbols +,-, and = rather than the words themselves. Also Boethius (475-526 AD) set up arithmetic for Roman numerals! Additionally, Gerbert of Aurillac, the future Pope Sylvester II, went to Barcelona to study astronomy and mathematics. He returned to Rheims in 972 to teach, reviving the abacus for counting (still in Roman numerals). Fulbert, Bishop of Chartres from 1006 to 1028, set up the first cathedral school, re-establishing finally in the west the teacher, student chain of transmitting knowledge that has not been broken since then. He encouraged learning for its own sake and taught the use of the astrolabe so that astronomical measurements began once more in Europe. Gerard of Cremona, working in Toledo, translated more than 70 works from Arabic to Latin (c.1150) while, also in Toledo, Archbishop Raymond set up a school of translation. Boethius had translated a few of Aristotle's works on logic. They had been the only source of information on Aristotle in the west but now more works of Aristotle became available in translations of Arab commentaries. Universities opened in Paris, Oxford, Bologna and Montpellier in the 12th century. In the next century there were also universities in Padua, Naples, Salamanca and Cambridge.

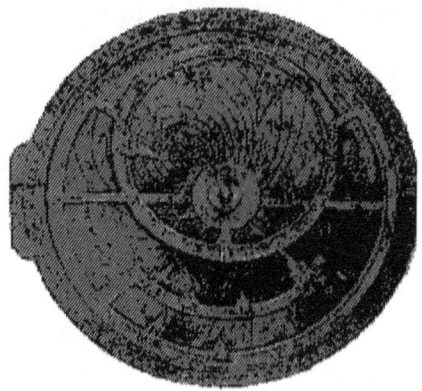

Figure 4. An Astrolabe

The Crusades finally renewed east-west contact. Though the contact was hostile, returning crusaders soon recognized and took advantage of the thirst for curiosities in Europe. Manuscripts as well as goods began to move from the Islamic to the Christian world and with the manuscripts came new ideas, especially the philosophy of Aristotle.

The Scholastics

Developments in Philosophy and Theology

Robert Grosseteste (1168-1253 AD), the Bishop of Lincoln, in England was persuaded by his study of Aristotle that scientific truth can be attained aside from revealed truth (truth revealed through Scriptures and church tradition). Scientific truth was to result from induction, that is, from the process of gathering information and then generalizing or drawing conclusions from that information. However, since generalizing from particulars can lead to erroneous conclusions if not enough information is used, scientific conclusions are only hypotheses that must be tested by comparison with experience (experiment).

Grosseteste's metaphysics were explicitly mathematical. To him, light was the basis of the universe and all causes were lines, angles and figures of light. Continuing these ideas and especially the emphasis on light, Grosseteste's follower Roger Bacon (1214-1294) drew upon a long tradition of optical theories in which mathematics played a major role. Because light travels in straight lines, Plato, Aristotle, Euclid and various Arab writers used ideas from geometry in discussing the behavior of light and the origins of vision. Bacon used mathematics to expand and elaborate Grosseteste's metaphysics of light.

Grosseteste was the first lecturer in the Franciscan school at Oxford and Bacon was an English Franciscan, the religious order that grew up out of the teaching and tradition of St. Francis of Assisi. St. Francis is famous for his positive attitude toward the world and the things in it. Because of this background, the Franciscans had a long tradition of tending toward experiment and observation as sources of knowledge. The Franciscans were especially dominant in England at Oxford University so it is no accident that the English came to dominate experimental science centuries later.

The many names of Albert the Great (Albertus Magnus or Albert, Count von Bollstad or Saint Albert of Cologne, 1206-1280) reflect the many facets of this great scholar. He began the task of reconciling Aristotelianism with Christianity. His student Saint Thomas Aquinas (1225-1274) completed the reconciliation in his famous and enormously influential *Summa Theologiae*. Eventually, this work became the official theology of Catholicism (it still is). The effect was to make both the advantages and disadvantages of Aristotelianism much weightier in the thought of succeeding scholars. In particular, by tying theology to a faulty Aristotelian science, Aquinas set theology and science on a collision course.

Shortly after the death of Aquinas in 1277, Pope John XXI ordered an investigation of doctrines taught at the University of Paris. The investigation, conducted by Etienne Tempier, Bishop of Paris, resulted in the condemnation of 219 propositions (along with excommunication for those holding them). Fundamentally, the Condemnation of 1277 was aimed at the "Doctrine of Double Truth" (Averroeism) under which (it was claimed) certain scholars believed both Aristotelian ideas and Christian ideas even when the two were directly opposed and contradictory. For example, one of the condemned propositions was the Aristotelian idea that the elements are eternal. This directly contradicts the Christian assertion that they (and the world) are created.

The Condemnation led to an eclipse of the work of Aquinas that lasted some 50 years. The Condemnation thus forced a reevaluation of Aristotelian science in light of Christian presuppositions. The physicist and historian of science Pierre Duhem called the Condemnation of 1277 the "birthday of modern science" because it led to explicit rejection of belief in the divinity of heavenly bodies. This opened the way to acceptance of the view that the heavenly bodies are independent objects that can be studied objectively.

Partly in reaction to the Condemnation of 1277, William of Ockham (Occam) (1280-1349) founded nominalism, the view that abstractions are only names and have no concrete reality of their own. For example, the word "mankind" refers to all people generally but only individual people are concretely real. There is no such object as "mankind"; it is an abstraction. Ockham believed that science should be a series of statements about real, individual objects. Generalizations about sets of objects are abstractions and cannot be as certain as statements about individual, real objects. Hence abstract statements are hypotheses which must be held with some skepticism. This is the Ockham of the famous Ockham's Razor (the Law of Parsimony) which requires us to use the simplest explanation consistent with the object of the explanation.

The overall impact of Aquinas' thought on the use of mathematics was neutral. It neither hindered nor advanced application of mathematics directly. Nonetheless, it laid the groundwork for the great changes to come from which modern science, with its characteristic impulse to apply mathematics to the world, would arise. Paradoxically, it also became the basis of later opposition to the new science. Ockham's nominalism, also mathematically neutral, was to play a simpler and more distinctly positive role in the rise of modern science. Though not an advance in applying mathematics, nominalism was a decided advance in the direction of using observational, particular language. This tendency would be a second important feature of the new science.

Developments in Mathematics

In 1220 Leonardo Fibonacci, who is still remembered for the Fibonacci numbers, published *The Applications of Geometry* and a book popularizing the use of Arabic numerals, *Liber Abbaci*. He was the best mathematician of his day as he proved by being the undefeated champion of the numerous scholarly competitions popular at that time in the courts of the Italian aristocracy. Around 1202, he introduced decimals to Italy. They were received with a little interest and considerable dismay. People were comfortable with Roman numerals and the abacus. Nonetheless Fibonacci's works contributed to the beginnings of a renewed interest in mathematics.

Four scholars at Merton College, Oxford in the early 1300's, Thomas Bradwardine, William Heytesbury, Richard Swineshead and John Dumbleton, greatly extended the mathematical descriptions of motion. In particular, the Condemnation drove the Mertonians to consider how things impossible to Aristotle could at least be rationally conceived. They did this by trying to show that the existence of such things (or the occurrence of these events) did not lead to logical inconsistencies. They found themselves forced to define terms carefully and this effort led to the first clear definitions of instantaneous velocity, uniform motion and uniform acceleration. Their greatest achievement was the Mean Value Theorem (which was long called the Merton Rule). In modern terms, they proved that if an object could be moved with constant acceleration, it would traverse the same distance in a given amount of time as a similar object traveling at a constant speed equal to the mean speed (the speed at mid-time) of the accelerated object for the same amount of time. For example, if a falling object happens to travel 10 ft in a particular second of its fall, the instantaneous speed equals the average speed (10 ft/s) at exactly half a second into that second of fall.

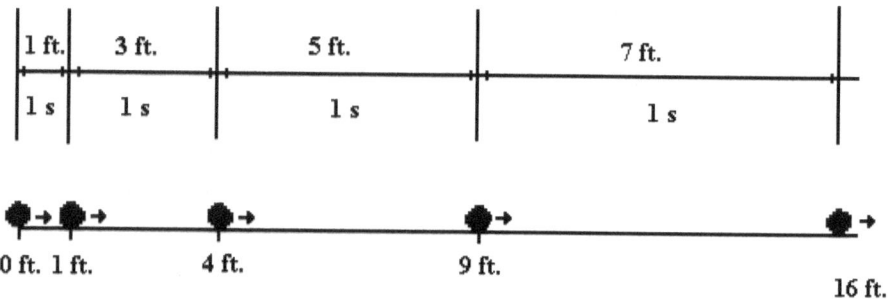

Figure 5. Uniform Acceleration. The object travels odd multiples of the first distance in succeeding time intervals.

Finally, Nicole (Nicolas) Oresme (1320-82) at the University of Paris developed the idea of representing qualities graphically although he did not use scales, coordinates or numbers but only geometric relationships. He found the shapes of the lines produced made motion easier to visualize and categorize. Representing velocities graphically made it an easy matter for him to prove the Mean Value Theorem geometrically. Furthermore, he was also able to prove that for a uniformly accelerated object, the distances traversed in equal, successive time intervals are odd multiples of the first distance, 1,3,5,7 etc. (Figure 5.). This result was later very helpful to Galileo in recognizing free fall as a uniformly accelerated motion.

The Decline of Scholasticism

By about 1400, the scholastic effort was running down. The Condemnation had stimulated efforts to learn and assimilate Aristotelian ideas but the threat of excommunication attendant on it encouraged philosophizing "according to the imagination", an attitude the pragmatic merchants thought sterile. Perhaps more significantly, the scholastic project was completed. The scholastics had set out to explore how Aristotelian ideas could be made consistent with Christian thought. Such tasks may be open-ended in principle, but in actual practice a point is eventually reached where new ideas are hard to come by and progress slows considerably. The schoolmen did produce, through nominalism, a skepticism about the validity of pure, deductive reasoning that pointed to the need for experiment.

The last major schoolman was Cardinal Nicholas of Cusa (1401-1464). Nicholas went back to Platonic ideas of using mathematics in metaphysics, moving away from the qualitative Aristotelian ideas. Cusa argued that our reason is limited and we are in a state of "learned ignorance". Though claiming extreme skepticism about our certainty of knowledge, Nicholas produced a cosmology where the Earth moves and is not at the center of the universe (there is no physical center). The overall impact of Cusa's efforts was a further weakening of confidence in the current knowledge. Thus, the desire for new ways of gaining valid knowledge was fed by Cusa's skepticism.

The Renaissance and the Reformation

The ferment begun by the East-West contact of the Crusades gave rise to great changes in the West. Scholasticism first and then the Renaissance and the Reformation sprang from the seeds sown by the Crusades. Major social and political changes accompanied these movements. The trade started by the Crusades accelerated and led to the rise of a merchant class, the middle class, that slowly undermined the feudalism of the Middle Ages. As the merchants gained influence, demand increased for people with better mathematical skills and for more calculational procedures to assist the merchants in their record keeping.

Rising trade needs drove exploration and discovery and, since this required accurate longitude and latitude information, clock making, cartography (the making of accurate, non-mythical maps) and mathematical and measurement skills became highly valued. Prince Henry the Navigator (1394-1460) set up his school for navigators in Portugal (of course, he had Portuguese dominance of the sea-lanes in mind). Accurate star charts were needed to guide merchant shipping. Responding, George Peurbach (1423-1461) and his student John Muller (a.k.a. Regiomontanus, 1436-1476) undertook the correction of the Toledo Tables, even going to Rome from Vienna to examine Greek copies of the *Almagest*. Tycho Brahe in Denmark began making new measurements of planetary and stellar positions. Nicolas Copernicus in Poland attempted to improve the Ptolemaic system and wound up creating an entirely new one.

Discovery undermined confidence in Aristotle and Ptolemy (and the Arab commentators through whom they were known). For example, Columbus' effort to sail west to the Orient was based on a 10th century error by the Arab astronomer Alfragamus who understated the circumference of the Earth (actually, the length of a degree at the equator) by 18%.

Columbus, believing the distance eastward to the Orient to be twice what it really is, arrived at a 3000-mile estimate for the westward distance to the Orient from Lisbon. The actual distance is about 14,000 miles! Fortunately for him, there was indeed a land mass 3000 miles west of Lisbon but it was one unknown to Aristotle and Ptolemy. More importantly, discoveries were made about Africa and other regions about which Aristotle might have been expected to have known. Hence, mercantilism increased the doubts about Aristotle originally engendered by the Condemnation of 1277. All this encouraged a search for other sources of ancient knowledge and ideas.

Humanism

The perceived sterility of scholasticism led some scholars to conclude that the "proper study of man is man" (and not God). Men like Erasmus and Sir Thomas More called themselves "humanists" to distinguish themselves from the futile schoolmen (there is in existence a letter where Erasmus greets Martin Luther as a fellow humanist). They turned their attention to society, law, and art and a search for other ancient sources and ideas began. Especially in Italy this produced the Renaissance. Some humanists, interested in power over nature, resurrected Greek and Hebrew mysticism and magic. In these magic systems, mathematics provides the key to control of the world because the underlying structures and harmonies of the world *are* mathematical. Thus, there was a beginning of a return to the Pythagorean belief that mathematics must be used if we are to understand the universe. In art, Leonardo da Vinci and others applied geometry extensively. Their success provided another bit of evidence to support the idea of the value of mathematics for describing, ordering and understanding the world.

Warfare

Social, political and religious change created an atmosphere in which warfare was almost continuous at a low level. The martial arts received attention from Simon Stevin, Leonardo da Vinci and others (later even Galileo). The advent of the cannon created needs for advances in mining, materials science, fortification and mechanics. Mining improvements were needed to find and develop old and new mineral resources for the manufacture of cannons, cannonballs and gunpowder. Advances in materials science were stimulated by the mining needs and by the need to improve the performance of the cannon. The dramatic destructiveness of the cannon forced abandonment of the high towered castle and stimulated a movement toward low fortifications of earthworks, a new design with new construction problems requiring reexamination of the fundamental principles of fortification architecture. Foremost in the mechanics of cannonball trajectories was Tartaglia (a nickname meaning "the stammerer", his real name seems to have been Niccolo Fontana-at least his will indicates he had a brother with that last name). Tartaglia (1500-1557), or Fontana, showed that the trajectories were very different from what Aristotle had thought. Studies of mechanics and fortifications were particularly mathematical so that warfare encouraged mathematical improvements and developments.

The military engineer Simon Stevin (1548-1620) of Bruges used the Archimedean method of exhaustion and, unlike the Greeks, was undisturbed by the idea of dividing a finite area into an infinite number of small areas. He also used irrational numbers, as did the humanist mathematician Gerolamo Cardano but negative numbers, long in use by merchants, were viewed as fictions by Cardano. Stevin, typically, had no qualms about negative numbers. Both he and Cardano used zero, a broadly accepted number by the mid-14th century, though their contemporary, Galileo, regarded neither zero nor one as numbers and distrusted algebra. Stevin advocated the use of decimals to replace fractions. Roman numerals had by now been displaced throughout much of Europe by the much more tractable Arabic numerals, making for an enormous improvement in computation times. Late in the period, about 1594, John Napier introduced logarithms to further facilitate calculations that were now becoming lengthy.

The Reformation

By about 1500, because of social changes and perceived corruption and inconsistency of the Roman Church, discontent with established sources of authority led to the Reformation with Luther, Calvin and Zwingli denying the religious authority of Rome and proclaiming that Scriptures alone are the source of religious authority. The great Reformation doctrine of the priesthood of the believer implied that the individual, guided by the Holy Spirit, was able to interpret Scriptures without relying on a priest as an authority. Education for the masses then became important and, indeed, Philip Melancthon, Luther's coworker, is credited with beginning the public school system in Germany. Note that mass education is impossible without the recently invented printing press. Ideas often must wait on technology for their enactment. Indeed, the printing press itself waited on the loom and the rise of the mercantile class. The loom produced a rag surplus that greatly increased the production of paper. Without surplus paper, the printing press would have been useless. The rise of the middle class produced a market for the large volume of books produced by the printing press. With Johannes Kepler, a member of that middle class and an early product of the new schools, began the rise of the new science, a science heavily dependent on mathematics.

There was a growing awareness, within Catholicism and especially within Reformation Protestantism that the creation is good and is created by a good and orderly God. This was consistent with the general Renaissance tendency to look for ancient sources of wisdom although the source in this case was the Hebrew Scriptures (Genesis 1) rather than Greek texts. In Catholicism,

St. Francis of Assisi (1182-1226) and his Franciscan order had began this emphasis, seeing in the world a reflection of its creator. For the Protestant Reformers, this idea was an underlying, little mentioned assumption permitting such changes as marriage of the clergy and the affirmation that all tasks and vocations have spiritual value. This doctrine of the Good Creation had a twofold impact on science. By its insistence on a rational, good Creator, it supported the effort to see unity in the chaotic world of appearances because it guaranteed order would be there. Also, because the Creator called creation "good", the doctrine countered the Greek revulsion for the world of appearances. Who dares call repulsive that which God has called good? In the hands of the English Puritans, this doctrine would have special importance for the development of English experimental science in the century following the Reformation.

The Renaissance and Reformation period created an intellectual climate much more favorable to science than any since the golden age of Greece. It freed men to think for themselves and provided new reasons for broad education of the masses. Furthermore, it provided the rationale for believing in a precise, mathematical order in the world (the orderly creation) and for believing the world of appearances could be experimented on, manipulated and made to reveal its secrets (the good creation). Without this encouragement, science in the west could not have arisen.

The Neo-Pythagoreans

Shortly before 1600 a number of men began to use numbers in areas they had not been used before and with a confidence in the value and productiveness of such undertakings not seen since the days of Pythagoras. Many of these men had been schooled at the University of Padua where a vigorous experimental tradition had long dominated the school of medicine. The nominalist and Franciscan traditions, dominant at universities such as Padua, Oxford and Erfurt (Luther's *alma mater*) had percolated throughout Europe by the time of the Reformation.

Also, by now many Europeans had a background of trade experience in using numbers for analyzing and assessing situations. Artists such as Leonardo da Vinci had shown that improvements in art and architecture could be made through the application of geometry and mathematics generally. Success gives rise to confidence but these successes alone are not sufficient to explain what was happening.

The Renaissance had made European scholars aware of the ancient Pythagorean program and had also led to a renewed hope in progress by human effort. This new optimism found support from Scriptures, as religious ages are compelled to do. The doctrine of the Good Creation, that the world was made by a good and orderly Creator, was central. The concatenation of all these influences, given sufficient time to germinate in the soil of the late Middle Ages, produced the Neo-Pythagoreans. Arguably, the doctrine of the Good Creation was the most important of these because, as we will see, it put backbone into the determinative work of Johannes Kepler. It also was a major shaper of the contemporary thought that would accept Kepler's work.

Advances in Medicine

A number of important medical advances were made through the application of mathematics to human and animal physiology. About 1614, Sanctus Sanctorius (Latin) or Santorio Santorio (Italian) (1561-1636) at the University of Padua set out to study human metabolism (his own, in fact) in an analytical manner. He had an enormous balance scale built on which he sat and slept, measuring his intake of food and output of wastes. He was even able to detect the effects of perspiration (or so he claimed)! Later, in 1626 he used the thermometer (invented by Galileo in 1596) to measure normal human temperature for the first time. At about the same time, Alfonso Borelli (1608-1679), a friend of Galileo at Padua, used mathematics and diagrams to explain movement of animals. From computations of the strength of the pectoral muscles of men and birds, he was able to show that the story of Icarus and Daedalus is impossible. The human pectoral muscles, he calculated, are not strong enough to power flight.

The most striking example of how mathematical thinking could help clarify biological problems is the work of William Harvey, an Englishman trained in medicine by the famous Fabricius at the University of Padua. He discovered in 1619 that blood circulates around the body, pumped by the heart into the arteries, to the veins, and back to the heart. The critical information leading him to this conclusion came from his measurements of the flow rate of the blood. He was able to calculate the weight of blood pumped by the heart each hour and he found it exceeded the total weight of the body! The calculation thus showed that blood must be recycled. In a concluding chapter of his book "On the Motion of the Heart", he remarked, "it has been shown by reason and experiment that blood by the beat of the ventricles flows through the lungs and heart and is pumped to the whole body".

Advances in Physics

"Johannes Kepler, Keppler, Khepler, Kheppler or Keplerus was conceived on 16 May, AD 1571, at 4.37 a.m., and was born on 27 December at 2.30 p.m., after a pregnancy lasting 224 days, 9 hours and 53 minutes. The five different ways of spelling his name are all his own, and so are the figures relating to conception, pregnancy and birth, recorded in a horoscope which he cast for himself. The contrast between his carelessness about his name and his extreme precision about dates reflects, from the very outset, a mind to whom all ultimate reality, the essence of religion, of truth and beauty, was contained in the language of numbers."[v]

Born into one of the oldest Protestant families in Germany, Kepler (1571-1630) was a life-long Lutheran although his local pastor briefly excommunicated him for refusing to profess the doctrine of ubiquity, an obscure doctrine later dropped by the Lutheran church. In his lifetime he was famous for his study of conic sections. This expertise stood him in good stead in his later investigations of planetary orbits that, as it turns out, can only have the shapes of conic sections.

From early student years, Kepler was a convinced Copernican. He was also absolutely certain that the creation of a good and rational Creator must necessarily follow strict and mathematically precise laws. His whole life was directed to discovering the mathematical forms of those mathematical laws. He succeeded to an unusual degree.

His first success was almost quixotic. While lecturing, he was struck by the idea that the sizes of the orbits of the six Copernican planets (Mercury, Venus, Earth, Mars, Jupiter and Saturn) about the Sun might be dictated by the fact that there are five Platonic solids. By nesting the solids within spheres and the spheres within the solids (Figure 6.), one can generate a set of concentric spheres whose relative size is controlled only by the order and sizes of the Platonic solids. There are 120 possible orderings of the solids but Kepler quickly found one that worked, almost. With a bit of cheating, allowing the spheres to have thickness, Kepler had his explanation of the sizes of the orbits. Moreover, he also had an explanation of why there were only six planets! It was all nonsense (as the discovery of more planets later showed) but he never seems to have realized it. More importantly, this seeming success, which made his name, encouraged his belief in the mathematical order of the universe.

This belief, and the success he attained following it, set the pattern of using mathematics for modern science. In particular, Kepler used this belief as a filter by which he sorted acceptable and unacceptable ideas. At one point in his work with Tycho Brahe's data, trying to find the exact shape of the orbit of Mars, he found that no combination of circles could fit the data to better than 8' of arc.

Knowing that the data was accurate to 4' of arc, he rejected circles and set out to find the true form. It was the first time in history a theory had been rejected because it did not exactly fit the facts. And the rejection made success possible. At the time, Kepler held the most important mathematical post in Europe (Imperial Astrologer) and the influence of this decision can hardly be overestimated. For this reason, Koestler has called Kepler "the watershed". Once Kepler had made his decision, science followed him and, by following him, was irrevocably changed. Eventually, Kepler found three mathematically precise laws of celestial (orbital) motion. We will examine them in detail when we look more closely at cosmology.

Galileo Galilei (1564-1642) believed quantification was the path to trustworthy and reproducible results. He wrote, "The book of the universe is written in a mathematical language whose alphabet consists of triangles, circles and geometric figures." The strong orientation toward geometry evident in this remark is no accident. Although Stevin's decimals were available, Galileo never used them nor did he ever use equations. He considered algebra and decimals inexact. He also had no clear idea of the continuum of numbers (the numbers line). In fact, many mathematicians in Galileo's day still followed Euclid in viewing the number 1 as the basic unit of numbers and not a number at all! These attitudes made it hard to think about continuous changes in motion (acceleration).

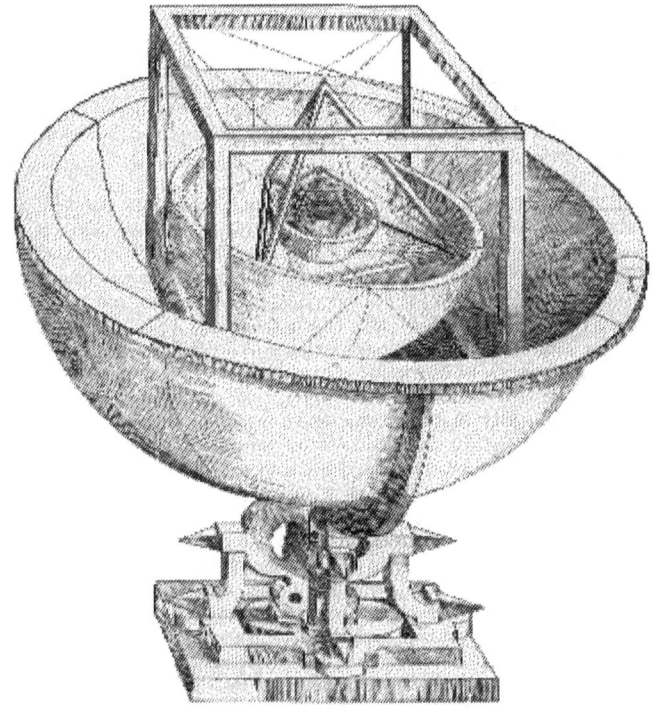

Figure 6. A model of Kepler's scheme for explaining the sizes of planetary orbits

Much more than a mere mathematician, Galileo relied heavily on observation and experiment. Hence, he is regarded as the first "modern scientist", someone who combined mathematics and experiment. Educated in medicine at the University of Pisa, he returned there at age 25 as professor of mathematics. He discovered that the pendulum swing takes the same time regardless of the size of the swing (the isochronous pendulum) and suggested its use in measuring the pulse of a patient.

Turning to physical motions and relying on Oresme's work, he showed that free fall motion is uniformly accelerated motion. Then he showed that projectile motion is simply a combination of a vertical free fall motion with a horizontal, constant velocity motion. Since Aristotle drew a radical distinction between free fall and projectile motion, this was a major conceptual advance. Galileo later became professor of mathematics at Padua.

In treating the planetary motions with mathematics, Kepler had been following tradition, not breaking with it. After all, even Aristotle believed the heavenly bodies move with mathematical precision. Galileo, by treating motions on Earth mathematically, was much more original. He was truly breaking with Aristotelian tradition.

René Descartes (1596-1650) revived the Pythagorean enterprise. In his "Discourse on Method", a preface to three sensational applications of his method in optics, geometry, and the science of meteors, he threw out the whole philosophic and theological tradition as useless and unproductive and unable to yield certainty. He proposed to replace them with a new philosophy built on the deductive model of mathematics.

Mathematics is a deductive structure requiring some initial propositions that are true (or assumed to be true), on which deduction can be applied to extract more truth. What was needed, Descartes decided, was a set of initial propositions that were certainly true. Then more could be deduced to build up a system of absolutely certain philosophic truth. Searching for such initial truths, he looked for propositions which were so clear and distinct that they could not possibly be doubted. He found he could not doubt his own existence ("I think, therefore, I am") and the existence of God. Building on these as initial axioms, he went on to attempt an explanation of literally everything. His system was very impressive, its influence almost too extensive to overstate.

In mathematics proper, Descartes invented the Cartesian coordinate system which made analytic geometry possible, reuniting numbers theory and geometry, areas of mathematics which had been sundered since the collapse of the Pythagorean brotherhood.

Kepler had shown the type of mathematical precision by which the heavens are run. Galileo had found precise mathematical descriptions of motions here on Earth and Harvey and Santorio had shown even the human body could usefully be studied mathematically. Descartes had created an imposing philosophical system based on the mathematical model. It remained for Isaac Newton (1642-1727) to construct a new, highly successful physical science. These developments led to a pervasive belief that all types of philosophical questions could eventually be answered with a mathematical certainty.

With the aid of Descartes' analytic geometry and Archimedes' method of exhaustion, Newton invented the differential calculus (simultaneously with the great German polymath Baron Gottfried Wilhelm von Leibniz) and the integral calculus based on the method of exhaustion. Using the calculus, he discovered three laws of motion and the law of universal gravitation. The combination of the laws and the new computational powers of the calculus enabled him to derive Kepler's laws and Galileo's results, unifying the sciences of heavenly and Earthly motions including comets and the tides.

He published his results in *The Mathematical Principles of Natural Philosophy*, (the *Principia*) setting as he did so the standard of deductive, mathematical science. Like Descartes, Newton organized his thinking with Euclidean precision. Definitions then axioms (the physical laws) then corollaries followed by lemmas, scholia and proved proposition after proved proposition march across its pages like a well drilled army. Newton first attacked the problem of the motion of bodies in free space (a vacuum). Successful in the first effort, he next assaulted the problem of motion in a resisting medium including motions in and of a fluid. Finally, he explained the "system of the world" including the motions of planets and comets and the behavior of the tides.

Reading the *Principia* was so taxing that John Locke (1632-1704) had to ask Christiaan Huygens (1629-1695) for reassurance on the mathematics since he himself could not understand it! Like Galileo, Newton did not use algebraic equations or decimals, trusting only the precise mathematical structure of geometry. All the proofs were geometrical and, hence, actually more awkward than modern versions would be.

Newton's followers were quick to use the new laws to chart actual planetary motions. Edmund Halley predicted the return of the comet now named for him and, eventually, the existence of Neptune was predicted (and later that of Pluto) based on Newton's physics. For more than two centuries, the new physics went from one success to another, encouraging many imitators and further strengthening the hope that mathematics provides the way to certain knowledge.

The calculus was considerably extended and advanced by the three great L's of French mathematical physics, Comte Joseph Louis de Lagrange (1736-1813), Adrien-Marie Legendre (1752-1833) and Marquis Pierre Simon de Laplace (1749-1827) and by Léonard Euler (1707-1783). All these men were skilled and original mathematicians interested in physical problems, especially celestial mechanics. They made the study and solution of astronomical problems much deeper and more mathematically subtle and elegant. Their work extended also to other problems almost too numerous to list. In the century and a half between the publication of the *Principia* and the death of the Legendre, Newton's work had been extended and transformed into a science of unprecedented depth and precision.

Newton himself became an intellectual hero of gigantic proportions, almost an intellectual deity. The attitude is wonderfully plain in Alexander Pope's "Epigram on Sir Isaac Newton",

> Nature and Nature's laws lay hid in night:
> God said, *Let Newton be*! and all was light.[vi]

Of course, there were dissenters. The disciples of Descartes had numerous complaints about features of the Newtonian system. We will consider them later. The calculus was itself not trouble free. The integral calculus (for finding areas and volumes) was a throwback to the method of exhaustion used by Archimedes and Kepler. Newton's technique involved dividing areas and line segments into infinitesimally small pieces and summing over all the infinitesimal pieces to find the total area or length. Many people found the idea of an infinitesimal piece of anything incoherent, which, of course, made the whole procedure incoherent. Bishop Berkeley was particularly critical of this aspect of the calculus, memorably labeling infinitesimals "ghosts of departed quantities". Bertrand Russell later judged the calculus "a tissue of lies."

Twentieth Century Developments

For more than two centuries, the application of mathematical procedures rolled from one triumph to another. Ultimately, confidence in the mathematical order of the universe comes from the successes of the idea. We tend to think, "It works, therefore it is true." Confidence in scientific methods grew without check. Then in the 20th century, two developments finally put limits on the method and undermined the previously boundless optimism in the method.

In Quantum Theory, the theory of atomically small objects, it was found that there are limits to what is knowable quantitatively. There are limits, set by the Heisenberg Uncertainty Principle, to how accurately certain quantities can be measured. The limits are very small but they are intrinsic and cannot be avoided.

Furthermore, individual events at the atomic level are unpredictable. Predictability appears only in the aggregate as many individual events are considered together so large objects composed of many atoms can behave in a deterministic way while the underlying behavior of the atoms is not deterministic. An actuary might be willing to predict 3 people out of 1000 will die next year but unwilling to predict, by name, exactly which particular people of any 1000 will die. Just so with atoms and subatomic particles, individual predictions are not possible. With people, it is conceivable that more information might make individual prediction possible. In the case of atomic and nuclear particles, there is no such possibility. Their individual behavior is inherently probabilistic.

Limits have also been found on the extent to which deduction can extract true conclusions from a system of "true" postulates. In the 1930's Kurt Gödel (1906-1978) proved two theorems that put limits on any mathematical or logical system large enough to encompass simple arithmetic. He first showed that there will be true statements in the system which cannot be shown to be true from within the system. With the second theorem he proved that the system cannot be used to prove its own logical consistency. The difficulties stem partly from the fact that some true statements require too many steps in proof. So the proof cannot be completed because no one has enough time to carry it out. There are also problems of the limitations of logic to decide an issue.

As an example of a truth that cannot be decided, Raymond Smullyan has considered the logical system consisting of all the statements one can make about an imaginary island that is inhabited only by knights who always tell the truth and knaves who never tell the truth.[vii] A visitor meets a native who says, "You will never believe I am a knight." By assumption, the native is either a knight or a knave yet, as the following argument shows, the visitor can never decide which. If the visitor should decide the native is a knight then the statement is false and the native is not a knight because he has not told the truth. Conversely, should the visitor take the native for a knave, the statement is true so the native must be a knight! Any decision by the visitor involves a contradiction within the system of statements. The decision cannot be made. But the native is either a knight or a knave although the truth of the matter is elusive. Thus, the deductive method can still lead to truth but it cannot lead to all truth. It cannot, in this case, decide the nature of the native. Hence, the Cartesian dream cannot be fully realized.

Interestingly, Gödel was much like Newton both in personality traits and interests. There are numerous legends of his tremendous seriousness, aloofness and absent-mindedness. Also, like Newton, he spent a considerable effort on theological concerns, in Gödel's case, particularly on attempts to prove the existence of God algebraically!

Unlike the discovery of irrational numbers, these developments have not led to the complete collapse of the enterprise. The application of mathematics to many types of questions has been too fruitful for absolute pessimism to arise. The unbridled optimism of the 18th and 19th centuries has, however, faded; very probably the new, more moderate, perspective will prove more realistic.

Further Reading

A History of Western Science, A.M. Alioto, Prentice-Hall, Inc., 1987.

Foundations of Modern Physical Science, G. Holton and D. Roller, Addison-Wesley Publishing Co., Inc., 1958.

Forever Undecided, Raymond Smullyan, Knopf, 1987.

Introduction to the History of Science, George Sarton, Carnegie Institute of Washington, 1927.

Seven Ideas that Shook the Universe, Nathan Spielberg and Bryon D. Anderson, John Wiley & Sons, 1985.

Science and the Making of the Modern World, John Marks, Heinemann Educational Books, Ltd., 1983.

Science since 1500, H.T. Pledge, Harper Torchbooks, 1959.

St. Francis of Assisi and Nature, Roger D. Sorrell, Oxford University Press, 1988.

The Beginnings of Western Science, David C. Lindberg, Univ. of Chicago Press, 1992.

The Discoverers, Daniel J. Boorstin, Random House, 1983.

The Sleepwalkers, Arthur Koestler, Grosset & Dunlap, 1959.

Chapter II

Experiment as a Valid Source
of Knowledge

The Classical Period

Classical Greek philosophical tradition was weak on experimental science. The Ionians and Pythagoreans relied on observation to a degree. Ionian speculations on the elements and Pythagorean musical studies were necessarily observation based to some extent. In contrast, the Platonic intellectual tradition was highly skeptical of, even antagonistic to, experiment for two major reasons. The collapse of the Pythagorean enterprise encouraged Plato to split reality into the world of perfect Platonic Ideals and the imperfect world of appearances. Appearances were seen as chaotic and confusing so studying them would lead to confusion rather than to reliable results. Additionally, Plato and his students were members of the leisure class for whom manual labor and dealings with the "nitty gritty" of everyday life was slaves' work unfit for a gentleman. Perhaps, there were class reasons for the Platonic repugnance for experiment.

Aristotle probably shared Plato's class attitudes but seems never to have liked the Platonic ideals. He was a careful and thoughtful observer of the world and, in fact, has to be rated as the first taxonomist. Alexander the Great, a former student, is said to have sent animal skins and dried plants to Aristotle for classification in the course of his conquests. Additionally, Aristotle had a family background in medicine with its necessarily "hands-on" approach.

Plato's depiction of the death of Socrates in his ***Phaedo*** set the stage for Aristotle's work. Socrates' students used Anaxagorus' materialistic physics to try to persuade Socrates to accept exile. In reply, Socrates said that a new physics is needed, one that takes into account purpose and ethical questions. Aristotle responded to the challenge, producing a new, and ultimately very wrong-headed, physics. His view of a world striving for perfection made it possible for him to think about the world of chaotic appearances as becoming, or trying to become, less chaotic and more ordered. He saw purpose in the changes of the world of appearances and purpose is a type of order.

However, Aristotle's works were lost to the West while Plato's were not. Fortunately, knowledge of Archimedes' very observation and experiment-oriented work was passed on to the West through Plutarch, so one example was not lost. However, detailed information on Archimedes' work was not available until 1543 when his surviving writings were translated into Latin.

It was in the Greek medical tradition where experiment first became a trusted method of learning how things really work. Hippocrates of Cos (460-379 BC), whose Hippocratic Oath is still the standard for medical practice, is the chief name associated with this tradition but towards the end of the era when Alexandria was the intellectual center, physicians such as Galen, Celsus, Herophilus and Erasistratus made important contributions to anatomy through the use of dissection. Of course, in an age lacking anesthesia, this was really vivisection performed on living criminals supplied by the state. For this reason, Tertullian called Herophilus a "butcher" who loved knowledge about men but hated men.

Galen (b. about 129 AD) did not dissect men but so extensive was his knowledge of animal anatomy and so voluminous were his writings that he dominated anatomy for more than a millennium. The word *experimentum* became, in the Middle Ages, virtually the synonym of "medicine".

The writings of Claudius Ptolemy played just as significant a role in encouraging experiment as a means of coming to understand the world as they did in encouraging the application of mathematics to understanding the universe. More in his *Optics* than in the *Almagest*, Ptolemy made explicit use of experiments in finding answers to questions about the world around us. He even went so far as to design special equipment for the sole purpose of doing experiments to establish his theories of exactly how reflected and refracted rays of light behave. This work was particularly important to Roger Bacon and to Witelo (c.1232-c.1275, no first name established). Witelo depended heavily on Ptolemy's *Optics* in preparing his *Perspectiva*, the standard optics text of the late Middle Ages. We know little of Witelo's life but we do know he, too, studied at the University of Padua.

The Interlude

During the Dark Ages and the Middle Ages, medicine was the chief repository of the experimental tradition. Unfortunately, observation was mostly a process of seeing what one knew was supposed to be there and, so, little new knowledge was added.

In the area of Astronomy/astrology, some observation was done. Ptolemy in the 2nd century and Copernicus in the 16th century were the only ones who seem to have done it for scientific purposes, however. Observation for predicting horoscopes and for setting dates of holidays, though it helped to keep technical knowledge of the needed measurement methods alive, did nothing to advance scientific study. Alchemy too helped retain some sense of the experimental method, although following the recipes and formulas of ancient texts was far more common than any real attempt to add to knowledge or break new ground.

On the negative side, a bizarre religious movement of the early Christian era added reinforcement to the Platonic revulsion for the real, physical world. Gnosticism, an amalgam of Platonism, Zoroastrianism, Judaism and Christianity, advanced a radical dualism between body and spirit, God and the physical world. Matter was evil, too evil even to have been created by God. Only spiritual things were ultimately important. The overall effect was to increase suspicion of the physical world and experimental investigation of it.

There were a few bright spots. Albertus Magnus, supplying what he explicitly saw as a deficiency in the ancients, wrote a ***Book of Minerals*** as the first attempt to put mineralogy on a systematic footing. It contained much information that simply had to be the personal observations of Albert. Along the way he gave some delightful and still relevant advice to would be alchemists: be patient, stay away from politicians and start with plenty of money.

The works of Alhazen (965-1040) came into the West around 1200 AD and substantially influenced Roger Bacon and others through him. Alhazen was the western name of the Arab scholar Ali al-Hasan Ibn al-Haytham who was one of the best scientists of the Islamic world. His main area of interest was vision and optical theory but he was no armchair theoretician. Following Ptolemy's example, he conducted his optical investigations using sighting tubes, strings (for tracing straight lines) and darkened chambers and candles. Additionally, he designed, made and used a number of instruments for measuring the angles of refracted light rays.

The synthesis of Aristotelian philosophy and Christian theology wrought by Aquinas encouraged the West to follow Aristotle by using observation more. Platonic idealism was thus increasingly neutralized just as the Franciscan emphasis on the good in creation began to undermine gnostic dualism. Both developments allowed more latitude in experimental work.

Though they seem to have done little experimentation personally, Grosseteste and especially Roger Bacon were vocal and influential advocates of experimental methods. Especially in Bacon's case, they bolstered their arguments with references to the work of the great Ptolemy and of Alhazen. They recommended experiment as a means of correcting error. Bacon was imprisoned 14 years to correct *this* error!

Theodoric of Freiberg used crystal spheres and an astrolabe to measure the refraction of sunlight and he recognized the rainbow as the result of such refraction in raindrops. He was even able to predict the inverted order of colors in the secondary rainbow from this work. Work of such exactitude helped enhance the growing perception of the value of experiment.

The prestige of experiment as means of gaining knowledge was also enhanced by the work of Pierre Pelerin de Maricourt. Publishing under the name Petrus Peregrinus, de Maricourt reported a series of clever experiments on magnetism in 1269. Using a spherically shaped piece of magnetic material, he discovered the poles of magnets.

Finally, William of Ockham and his nominalist followers, with their insistence that generalizations are hypotheses and that only statements about particular objects and events can be relied on, turned the attention of the late middle ages to questions of the validity of experimentally acquired information. Experiments are quintessentially tied to particulars.

The medical school of the University of Padua was the main bastion of experimental work in the late middle ages through the Renaissance and Reformation. A few students who left the University of Bologna and, taking their teachers with them, moved to Padua founded it in 1222. Italian universities of that era were basically concentrations of students who hired their own teachers. The students fired unpopular teachers. This type of university structure was short-lived. Different student groups, contending with each other to hire or fire favorites, soon became rioting mobs that attracted the attention of local authorities and popes alike (not always because of similar intentions since popes and local authorities were often at odds). Authorities became increasingly involved in regulating university activities. Student power necessarily waned.

The University flourished. By 1250 it had two chairs of medicine, three by 1262. It very early became known as a hotbed of Aristotelianism (which was soon called Averroeism in the wake of the controversy surrounding the Condemnation of 1277). Early in the fourteenth century, epidemiology had its beginnings as a medical science in Paduan studies of the plague. From this work sprang new sanitation and quarantine laws in Venice, Ragusa and other towns.

By the fourteenth century, dissection was permitted and began to be common in the medical schools. Surgery, until then more closely associated with the University of Bologna, became a major strength of the medical program in Padua. The great Vesalius, who had done dissection from boyhood as the son of a physician, wrote his famous treatise *The Fabric of the Human Body* while teaching at Padua. Other great names in this Paduan tradition are Giovanni Bonacci (who wrote under the name Gabrielle Falloppio), a student under Vesalius who advanced the knowledge of human anatomy well beyond that of Vesalius; his student Fabricius (Girolamo Fabrizio d'Acquapendente) who discovered the valves of the veins and taught William Harvey, the discoverer of blood circulation pumped by the heart. Galileo, trained in medicine at Pisa, spent the happiest and most productive years of his life as Professor of Mathematics at Padua.

Although his work was not published and had no impact on later thinkers, Leonardo da Vinci is a superb example of the new attitude toward experiment characteristic of the Renaissance. da Vinci's fascination with and designs for a variety of different gadgets, including flying machines, are well known now though few of his contemporaries were aware of his activities in this area. Leonardo also dissected human bodies and formed theories about the action of the eye in vision and the function of the heart. He helped in design of fortifications and did studies of the trajectories of cannonballs.

Experiment-The New Tool

The slow advance of the fortunes of experimental science during the Middle Ages and Renaissance produced a growing tension that, like building stress underground, would erupt like an earthquake around 1600. The efforts of Tycho Brahe, Kepler and Galileo were the first great tremors of the quake that has permanently altered our intellectual terrain.

A Danish noble, Tycho Brahe (1546-1601) was impressed as a teenager with the inaccuracies in the current planetary tables. He became obsessed with the need to improve knowledge of planetary positions and spent his entire adult life pursuing that goal. With the financial support of his king he was able to build the finest non-optical sighting instruments ever attempted (Figure 7.) and with them he produced the best data available on planetary positions before the advent of the telescope.

It was to Tycho that Kepler had to go to get the data he needed to discover his laws, and it was to Tycho's accuracy that Kepler bowed when he refused to continue working with circles to describe the motions. Tycho also observed the 1572 Nova (now called Tycho's Nova) and by his incomparable measurements was able to persuade his contemporaries that the "new star" was immobile and in all respects star-like. He also measured the distance to the Great Comet of 1577 and showed it had to be a considerable distance beyond the Moon.

These observations defied the Aristotelian doctrine of the immutability of the heavens, the view that the only changes in the heavens were due to the circular movements of heavenly bodies. Neither cometary motion nor the brightness changes of the Nova fitted this dogma. These results showed what a difference carefully controlled and measured observations could make in determining the truth or falsity of beliefs and propositions. Tycho had found a critical counter-example to Aristotelian ideas and no one could doubt that something was very wrong with the doctrine of the immutability of the heavens.

Galileo was both a mathematician and a careful experimenter. That is why he is regarded as "the first modern scientist". We have already discussed his experiments in free fall motion. He was also famous for his observations with the telescope. His observation that Venus has phases like the Moon destroyed the credibility of the Ptolemaic system and provided yet another example of the power new observations could have in determining the truth of various theories. With Tycho, Kepler, and Galileo, experiment finally had gained the recognition it deserved as a means of sorting out what is true from what is false.

Further support for experimental work came from William Gilbert, the personal physician to Elizabeth I of England. Another Padua trained scientist, Gilbert labored for some seventeen years studying magnetism. His account of his efforts, **De Magnete**, was published in 1600, just three years before his death of the plague. It is a sober, careful work with details of many experiments. Both his methods and his results greatly impressed contemporaries and the book was enormously influential, leaving an imprint on the explosion of scientific developments of the seventeenth century. Gilbert stated his view of the value of experiments in the first lines of the preface to the book, "stronger reasons are obtained from sure experiments and demonstrated arguments than from probable conjectures and the opinions of philosophical speculation"[viii]

Figure 7. Tycho Brahe at work in his observatory. Note that even in the sixteenth century, good science was expensive.

What Descartes was to mathematics, Francis Bacon (1560-1626) was to experiment. Bacon concluded that most of the previous efforts of the Greeks and the Scholastics were futile. He thought experimental efforts so far had been ill conceived and haphazard. What was needed, he believed, was a new experimental method. He called it the new tool ("novum organum"), recalling Aristotle's naming of logic as the tool of science. He had in mind a very systematic method with careful record keeping aimed at the acquisition of a vast amount of information from which generalizations later could be extracted. Data acquisition was to be carried out before theories were developed (he feared prior theorizing would prejudice the data taking). He called for discussions among the various experimentalists in different areas so that cross-fertilization could result in an interweaving of knowledge.

Bacon believed a gain in knowledge would result from the new connections thus uncovered. The method would also avoid the errors of the past, the "idols of the mind" which Bacon carefully categorized. His discussion of these errors was historically important because it enabled people to recognize and avoid them. Idols of the den are those individual biases we have due to background and experience. Moving outward from ourselves, the idols of the tribe are the biases we have because "all the perceptions both of the senses and of the mind bear reference to man and not the universe". Moving into society, the idols of the market come from the dominant views of our times and idols of the theater come from the dogmas of accepted religious and philosophic systems, from the great world of historical events and major movements.

Fittingly, Bacon died as the result of an experiment. In the winter of 1626, it occurred to him to see if snow could be used to refrigerate chicken. While stuffing the carcass of a hen with snow, he reportedly caught a chill that developed into a fatal case of bronchitis.

The Stewards of Creation

More than any other single factor, the group of men who began to meet informally in London in the 1640's and who organized officially as the Royal Society under the royal patronage of Charles II in 1662 were the source of the new experimental science. There had been scientific societies before. Galileo was so pleased with his membership in the Society of the Lynx that he continually referred to himself as the "Lincean professor" in his *Dialogue Concerning the two great World Systems*. But the Royal Society was formed in a different environment and with a new emphasis on encouraging experiment and invention.

There was no Inquisition in England to dampen intellectual investigation and discussion and the mid-17th century was a time of great Puritan influence in England. The Puritan saw nature as a secondary revelation of God so that the study of nature was a proper function of man as the steward of the created world. Though not a Puritan, even Francis Bacon did not escape their influence. The first line of his *Novum Organum* is "Man, as the minister and interpreter of nature…". Science had an important role to play because a steward needs to understand what he is charged to care for. Scientific results were expected to benefit society and that expectation provided both a rational for scientific activity and an incentive for it that even the monarch approved. Puritan educational reforms, driven by the doctrine of the priesthood of the believer and by the egalitarian ideas implicit in that doctrine, encouraged mechanical sciences.

Reading the works of many of the member of the early Society, one finds explicit references to the ways the new science glorified God. As a group, they were devout Christians and seem to have thought of themselves as defenders of the faith against the atheists such as, they supposed, the political theorist Thomas Hobbes (1588-1679).

Robert Hooke, Robert Boyle, Christopher Wren and Edmund Halley were early members. Newton was elected to membership in 1672 for his invention of the Newtonian telescope. Charles II was the society's patron but Francis Bacon was its patron saint. Bacon's influence on the society is well documented. In fact, Robert Boyle was so Baconian he long refused even to read Bacon's **Novum Organum** for fear he might be "seduced too early by lofty hypotheses"!

It is instructive to note some of the important early members and their experimental work. Hooke discovered the law now named for him relating the extension of a spring to the extending force. Hooke's Law is often written: $F = kx$ where F is a force applied to a spring and x is the resulting stretch of the spring. The spring constant, k, is an indication of the stiffness of the spring. Hooke may have provided some necessary insight for Newton in the discovery of the law of gravitation. Robert Boyle was a prolific experimenter and a founder of modern chemistry. His experiment records fill many volumes. He discovered Boyle's law of gases (known also as Mariotte's law in continental Europe). That is, Boyle found the pressure of a gas is inversely proportional to the gas volume (at constant temperature) so that PV = constant. Boyle also led "The Company for Propagation of the Gospel in New England and parts adjacent in America", a mission society supporting missions and translation of the Bible into Indian languages.

Newton was a gifted experimenter, especially in optics and color. One of the most familiar paintings of Newton, by John-Adams Houston, shows him in a darkened room, demonstrating the ability of a prism to refract sunlight into a rainbow of color. The painting dramatizes but does not exaggerate Newton's experimental side. By passing the colored rays through three prisms, breaking, recombining, then again breaking the white light into the colors of the rainbow, he showed convincingly that color is an inherent characteristic of the light and not a characteristic imposed on the light by the prism. Newton later called this his *experimentum crucis*, the crucial experiment in understanding light. Religiously, the greater part of Newton's intellectual efforts was devoted to theological speculations rather than science.

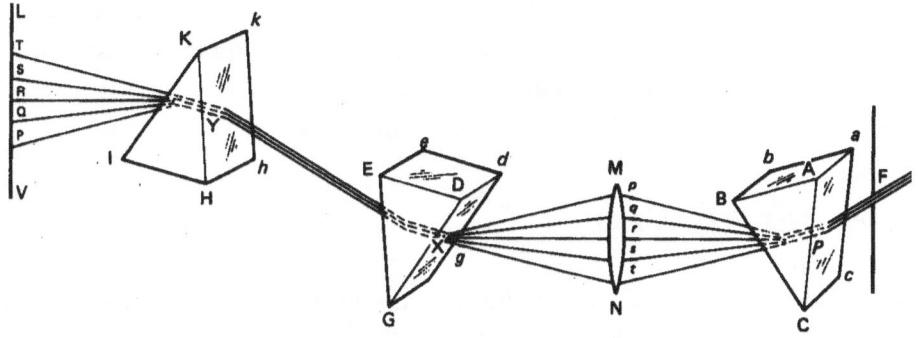

*Figure 8. Newton's **experimentum crucis**. Light enters on the right. The first prism breaks the white light into colors. A lens breaks the colored rays together and they are recombined to white light by a second prism. A third prism then once again breaks the light into colors.*

Newton's *The Mathematical Principles of Natural Philosophy* (usually just referred to as "the *Principia*"), in Latin and rigidly deductive in method, is (and was for Newton) the prime example of precise physical science. His *Opticks*, in English and written so as to stress experimental evidence, was the standard of good experimental science for years. John Locke, Benjamin Franklin and others who could not follow the mathematics of the *Principia* learned scientific method from the *Opticks*. Evidently Newton wrote the *Opticks* in the style he did out of despair of being able to express his optical knowledge with the precision he had been able to apply to the study of motion. Of course, the *Opticks* is still quantitative but it is not as thoroughly and mathematically deductive as Newton probably wanted it to be.

The successes of a few people, Tycho, Kepler, Galileo, the English experimentalists and especially of Newton who, following Galileo, combined experiment with mathematical theories, turned the suspect experimental method into an accepted means of finding truth. The doctrine of the Good Creation may have given the initial encouragement but, eventually, success is the most persuasive argument for belief. Experiment became an acceptable source of valid knowledge and the experimental method passed into the mainstream of science. The centuries since then have served only to entrench it more firmly as the major feature of science.

Scientific Methods

The classical Greeks set a deductive standard for scholarship that was followed even, at times, slavishly until the sixteenth century. The mathematics first of Thales and later of Euclid provided the model for clear reasoning. In a deductive system one begins with some, presumably undeniable, truths as the axioms of thought and generates more true statements from the axioms by the rules of deduction. Especially in the physical sciences, deduction is largely carried out in mathematical form. Deduction in verbal form is used in all the sciences. That is why precisely defined terms and correct language are extremely important in all the sciences. Deduction can only be used on sharply defined terms. We have seen the deductive proofs that $\sqrt{2}$ is irrational and that the area of a circle is πr^2. The simplest example of deduction is the Aristotelian syllogism. It consists of two premise statements from which a conclusion statement then follows rigorously. For example, if the premises are 1) All professors are intelligent and 2) Dr. Cramer is a professor, then the conclusion must be that Dr. Cramer is intelligent. The attractive feature of deduction is that the conclusions are inescapable. In fact, the word "proof" only properly applies to deductive arguments and only the conclusions of deductive logic are rightly described as proved.

The weaknesses of the procedure are twofold. Firstly, mistakes can be made in the drawing of conclusions. Mistakes are increasingly likely as arguments become longer but even with the short Aristotelian syllogism erroneous conclusions can arise. If nothing is better than a juicy steak and moldy bread is better than nothing, does it follow that moldy bread is better than a juicy steak? Secondly and more problematically, the axioms or premises may not be true after all. As one might suspect, axioms that are certainly true are not all that easily found. Descartes' second axiom, that God exists, has been doubted although he proclaimed it too clear and distinct to be doubted. It may even be possible to doubt his first axiom (that he existed).

Axioms have come from a variety of sources. Speculation, observation, revelation and (at least in some forms of mathematics) pure imagination. Increasingly, the deductive process itself has become symbolic (mathematical). As experimentation and observation became important with Tycho, Kepler, Galileo and Newton a shift occurred away from the other sources. It thus became characteristic of the physical sciences that observation and experimentation provided the axioms from which conclusion were deduced mathematically.

The actual process of coming to a scientific understanding of a phenomenon was and is more complex than a pure deductive model permits. For example, the conclusions (or predictions) of a scientific deduction are also checked by observation and experiment. If experimental evidence tells against the conclusion, the two weaknesses of deduction must be reexamined. Was an error made in the mathematics? Even as great a scientist as Einstein occasionally made mathematical mistakes.

The premises of the deduction will also need to be reconsidered. At this point the whole question of how premises are formed from experimental or observational information inevitably arises. Premises are generalizations. Experimental information is particular information in the sense that it pertains to particular examples or cases. For example, if examination of a sample of fish in a lake turns up a 37% infestation rate of a certain parasite, we then conclude that 37% of fish in the lake are infested by the parasite. This conclusion might then become an axiom for the conclusion that the Department of Natural Resources must undertake a particular action. Actually, all that is certain is that 37% of the sample was infested.

Since all the fish in the lake were not examined, we do not know that 37% of them are infested. We have generalized from particulars.

Generalization from particulars is known as induction. It was this process that Francis Bacon advocated. In contrast, deduction moves from generalizations to particulars. The weakness of induction is that all particulars must be known if absolutely certain generalizations are to be possible. Usually, all particulars cannot be considered. Taking samples of a fish population is a tacit acknowledgement that examining all fish is just not practicable. Of course, there are good and bad sampling techniques. Not even good techniques, however, entirely escape this problem of induction.

Contrary to what you may have read or heard, there is no one scientific method. Scientific methods all involve both deduction and induction. Induction is used to check conclusions as well as to generate axioms. Deduction is used to make predictions and to interconnect ideas. There are no fixed rules for the order in which these processes operate. Bacon thought induction was fundamental and should be the first and prime process. Books claiming there is a single scientific method usually follow Bacon's model. This is a caricature of scientific procedures.

Because both deduction and induction require specialized procedures and techniques, there is division of labor in the physical sciences. Theorists specialize in deduction, experimentalists in data gathering. Theorists explain phenomena and try to predict new phenomena. Experimentalists study things experimentally and try to discover predicted new phenomena. Occasionally, an experimentalist has the good fortune to discover something unexpected.

Induction, however, is the prime process in science in the sense that experiment is the final authority. Plausible theories that explain many things may be proven wrong by just one, well-chosen experimental result. The demise of the Ptolemaic system after Galileo's discovery of the full phase of Venus is an example of this power of the counter-example. Thus, science is not relative. Ideas cannot be true for one scientist and not true for another. Unless the known facts are not determinative, ideas are either true to the facts or they are not. It is this characteristic that makes people call science "objective". More importantly, it is this characteristic that forces on scientists what Jacob Bronowski has called the "habit of truth". Without honesty about results, science would rapidly become a house of cards.

Further Reading

The Logic of Scientific Discovery, Karl Popper, Harper Torchbooks, 1959.

Science and Human Values, Jacob Bronowski,

Science in the Middle Ages, ed. David C. Lindberg, Univ. of Chicago press, 1978.

Science: Growth and Change, Henry W. Menard, Harvard Univ. Press. 1971.

Science: Men, Methods, Goals, ed. B.A. Brody and N. Capaldi, W.A.Benjamin, Inc., 1968.

Scientific Literacy and the Myth of the Scientific Method, H.H. Bauer, Univ. of Illinois Press, 1992.

Chapter III

Structure of the Cosmos

The Classical Period

The Ionians seem to have been first people to use geometric models of the universe. Such models are especially simple and comprehensible (to the mind trained in geometry) and are therefore more easily checked for the correctness of predictions. The Ionians advanced a variety of cosmological schemes in which the Earth was a flat disk and the stars were either fiery objects embedded in the sphere of the heavens or holes in that sphere through which the light of an external fire could be seen. Anaxagorus (500-428 BC) clearly recognized that in lunar eclipses the Earth shadows the Moon and in solar eclipses the Moon shadows the Earth. He also taught that the Moon shines only by light reflected from the Sun. He was expelled from the Athens of Pericles for teaching that the Sun is a red hot rock and the Moon is made of earth. Belief that heavenly bodies were gods hindered the development of science until the Condemnation of 1277 forced Western thinkers to recognize their status as mere objects.

The Pythagoreans agreed with Anaxagorus as to the Sun and Moon and they insisted on a spherical Earth. Idiosyncratically, they believed the Earth revolves around a central fire (located at the center of the universe) simultaneously with another planet called Counter-Earth which was always across the central fire from Earth and, hence, invisible. It is tempting to think the Sun and central fire were identical but the Pythagoreans distinguished the two. The Sun made night and day. It could be seen from anywhere on Earth (during the day). The central fire, however, was never visible from Greece (some Pythagoreans claimed it was visible in India!). Ten objects (the sacred number) circled the central fire.

The Pythagoreans regarded the circle as the most perfect geometric figure and believed all motions in the heavens were circular. It is worth remembering, however, that the Pythagoreans were not geocentrists. An Earth forever circling the central fire is surely not at the center of the system.

They were disgusted that mountains prevent Earth from being a perfect sphere. Indeed, this attitude long persisted. Casanova (1725-1798), when forced to travel through the Alps, rode the entire way with the curtains of his carriage drawn. He wanted no accidental view of the excrescencies that disfigure what should have been a perfectly spherical Earth. Our pleasure in mountain views is recent, an outgrowth of the European Romanticism of the nineteenth century

Plato's Problem

Plato believed understanding of the changing world of appearances would only come as one recognized the underlying ideal forms. Circular motion, which has no end or beginning, could continue forever at a constant distance from the center. It was one of the ideal forms. The stars, traveling in daily circles about the Earth, seemed to him to perfectly exemplify this idea and to suggest that the heavenly motions were perfect. The wanderers (planets, including the Sun and Moon) partook of this diurnal motion and so, Plato concluded, were part of the heavens. The heavenly motions appear to circle the Earth, so, trusting appearances at least this one time, Plato set the Earth at the center of his system. How, then, was the meandering aspect of their motion to be understood? Surely in terms of more intricate arrangements of circles than was required for the stars. Thus Plato was led to propose to his students what is now known as *Plato's Problem*. What is the combination of circles that accounts for the observed motions? Plato's ideas constitute the first elaborated geocentric system of the cosmos. In his *Timaeus*, Plato sketched out his idea of the order of the planets. "The moon he set in the orbit nearest the earth, the sun in the next and the morning star (Venus) and [Mercury]".[ix]

Probably the best mathematician at the Academy was Eudoxus. He gave the first answer to Plato's question, finding that 27 nested circles or spheres were necessary. The Sun and Moon each required 3, the other 5 planets (Mercury, Venus, Mars, Jupiter and Saturn) needed 4 each. The 3 or 4 spheres were nested, the smaller within the larger. The final sphere was that of the fixed stars whose daily rotation was shared by all the other spheres. From the outset it is unclear if all the stars were thought to be the same distance from the Earth. That is, the sphere of the stars may have been a shell. With no ability to measure such distances, the Greeks were uniformly silent on the subject. Eudoxus seems to have been unable to decide whether the spheres were mere logical constructs or real but invisible objects. This indecision was never fully resolved in all the centuries of belief in geocentrism.

Plato's view of the cosmos was animistic. That is, he thought of the universe as an animal or organism rather than as a mechanism. In the *Timaeus*, there is an amusing section where Plato explains that the universe has no eyes nor legs nor feet because it has no need of them! Plato may have thought the universe had a beginning but one cannot be entirely certain. Remarks in the *Timaeus* implying a beginning of the universe may be purely metaphorical.

Aristotelian Cosmology

Aristotle and Callippus of Cyzius were both moved to tinker with this scheme of Eudoxus. Callippus, attempting to correct deficiencies of the scheme with respect to the motions of Mars, Venus and Mercury, added 8 more spheres for a total of 35. To resolve problems relating to the presence of a medium between the planets, Aristotle then added 21 more and the total rose to 56.[x] Possibly as an extension of Plato's suggestions in the *Timaeus*, Aristotle thought the heavenly bodies were made of a fifth element, aether, the most perfect form of matter which was never found on Earth. The heavens began at the orbit of the Moon and were radically different from the imperfect Earthly region where things were made of the 4 ordinary elements: earth, water, air and fire.

Regarding the Earth, Aristotle noted 3 observations that show the Earth is spherical. The disappearance of a ship hull on the horizon while the mast remains visible is due to the rounded surface of the water intercepting the line of sight (Figure 9.). The reports by travelers that new stars appear and familiar northerly ones disappear as one travels south are accounted for by the same argument. Finally, the shadow of the Earth on the Moon during a lunar eclipse is curved and never straight.

Aristotle also had a proof that the Earth is spherical and another that the Earth cannot rotate. Both of these proofs relied on Aristotelian ideas of motion. Aristotle divided all change, motion included, into natural changes and those changes produced by an outside agent. The latter he called "violent" change. The natural motion of the element aether was obviously circular while that of water and earth was vertically downward and that of fire and air was vertically upward. Natural motions continued until their purpose was completed but violent motion ceased as soon as the agent stopped causing it. Since there are no agents in the heavens, only circular motions can occur there. This conclusion is known as the doctrine of the immutability of the heavens.

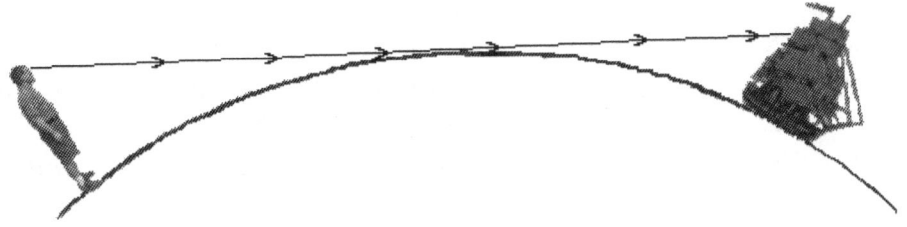

Figure 9. The rounded surface of the ocean intercepting the line of sight to a ship.

To prove the Earth is spherical, Aristotle imagined that at one time all the 4 elements were scattered through the universe. Since their natural motion is toward the center of the universe, earth and water soon collected at the center of gravity. Later arriving pieces filled any gaps in spherical shape because they tried to get as close as possible to the center. Hence, the Earth was built up in a spherical fashion. If the Earth were to rotate, that would be a violent motion of a very large body and would require the continuous application of a large force. Since no such force or agent exists, the Earth cannot rotate.

As the previous discussion demonstrates, Aristotle used deduction and observation together in a manner reminiscent of modern science. In fact, we regard his views on nature as the first example of scientific thinking and refer to them as "Aristotelian science". However, his view of the universe, like that of Plato and unlike that of modern science, was animistic. He viewed the universe as a sensitive organism with purpose and emotions. Also, to Aristotle, the universe was finite although, because of the lack of stellar parallax, he knew the size of the universe was enormous compared with that of the Earth. He considered the universe eternal.

A Dissenting Minority

Not everyone assented to these schemes. Though it would soon wither away, dissent came from several distinguished sources. Picking up the atomism of Democritus, Epicurus of Samos (341-270 BC) argued that the idea of a boundary to the universe was an incoherent idea and a boundary could not exist. Since this implied an infinite universe, Epicurus denied there is a center to the universe. The later Epicurean, Lucretius, indicated Epicurus thought there were more worlds than ours with other animals and people.

Heraclides of Pontus, a contemporary of Aristotle, pointed out that the daily motion of the stars could just as easily be due to the rotation of the Earth and remarked that, to an observer off the rotating Earth, an arrow shot vertically up would appear to be a projectile. Since Aristotle viewed vertical motions as totally unrelated to projectile motion, this provided an alternative to the Aristotelian position.

It was well known that the planets Mercury and Venus always appear near the Sun. Scholars have speculated that Heraclides was the first to suggest they are like moons of the Sun, orbiting it rather than the Earth. However, it does not appear that the idea goes back that far.

Appolonius conceived the idea of setting the center of a small sphere (the epicycle) on the rim of a larger sphere (the deferent) rather than nesting the spheres as Eudoxus had done. He thereby made possible an improvement in geocentrism that Ptolemy would exploit.

Aristarchus of Samos (310-230 BC) invented ways of measuring the distances and sizes of the Moon and Sun. Probably he was aware of Phoenician exploration down the coast of Africa (perhaps they had even circumnavigated Africa by 600 BC). The sphericity of the Earth implied by this navigational feat and the nature of lunar eclipses indicated to him that the Sun is much larger than the Earth. A large Sun might well seem a more likely object for the center of the universe and that might well explain why Aristarchus proposed the first heliocentric system. It was not taken seriously by his contemporaries. Cleanthes the Stoic, in fact, accused him of impiety for putting in motion the "hearth of the heavens".

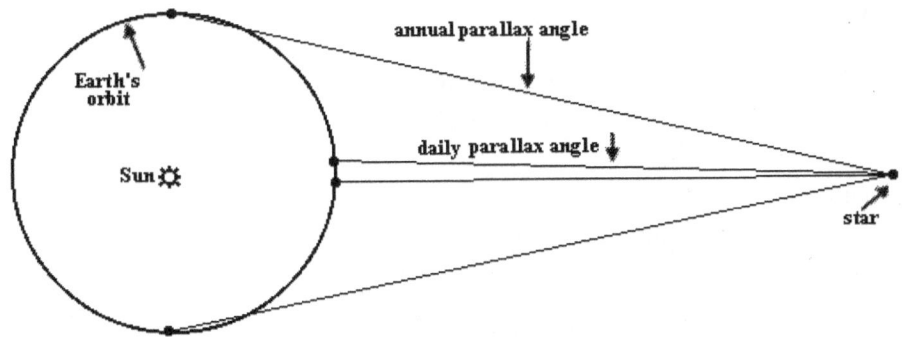

Figure 10. Annual and diurnal parallax of a star.

More serious was the problem that no annual parallax could be found (Figure 10). Parallax is the change in direction of the line of sight to an object as the object or the observer (or both) change position. For example, if you want to look steadily at an object as you walk past it, you will have to turn your head to see it. Your line of sight to the object changes so you must turn your head to keep your eyes looking in the right direction. Now the greater the distance to the object is compared with the distance of the motion, the smaller will be the parallax angle. Annual parallax is the change in the direction one must look to see a star or planet as the Earth moves on its orbit around the Sun. It must be much greater than daily parallax because the daily movement of the observer, due to the rotation of the Earth on its axis, is much smaller. No annual parallax was observed, implying a much greater distance to the stars than implied by the lack of daily parallax. This greater distance was unacceptable. Aristarchus' contemporaries preferred to think the whole idea of the motion of the Earth was nonsense. They wanted to believe the lack of annual parallax meant there was no annual motion of the Earth.

The Ptolemaic System

Working at Rhodes about 150 BC, Hipparchus improved on the methods of measuring distances to the planet, laying as he did so the foundations of trigonometry. He evidently had access to the ancient Babylonian records and was able to detect the precession of the equinoxes due, as we now know, to the slow precession of the rotational axis of the Earth. The long observational period from the Babylonians to Hipparchus was necessary because the precession is very slow with a period of 26,000 years. His calculation of the distance to the Moon was quite accurate. He found it to be 59 times the radius of the Earth, a correct result from the surface of the Earth to the center of the Moon but incorrect for a from center to center value for which the right factor is 60 times. Another very impressive achievement of this great astronomer was his calculation of the lunar month to within one second of the modern value!

Hipparchus was best known in antiquity for the new observations of planetary positions he made from his observatory on Rhodes. These observations did not fit earlier geocentric schemes and compelled the construction of a new scheme by Ptolemy. The Ptolemaic system constructed at Alexandria about 150 AD by Claudius Ptolemy relied on all the previous work although Ptolemy added his own observations to those of Hipparchus and the Babylonians. Ptolemy may also have been the first person to identify position on the Earth by longitude and latitude.

Rejecting the nested spheres of Eudoxus, Ptolemy worked extensively with the epicycles of Appolonius and with eccentrics, a variation of the epicycles first used by Hipparchus. Additionally, Ptolemy used a most peculiar invention of his own, the equant (Figure 11). Epicycles and eccentrics were primarily orbit shaping devices while the equant adjusted planetary speeds on the orbit. In Figure 11, the center the eccentric circle is not at the center of the Earth. In the equant device, the equant point is a point, not at the center of the Earth, from which the motion of the planet supposedly appeared constant. The small circles are the epicycles proper and the larger circles were called deferents. The equant seems to have been a salve to Ptolemy's conscience, a means of retaining perfection (constancy) while also eliminating it.

His system, expounded in his book *He Mathematike Syntaxis* (called the *Almagest* by the Arabs), required a total of only about 15 spheres. That figure reportedly inflated to as many as 80 in the Middle Ages to account for deviations of observed positions from predicted positions. His attitude toward the question of the real existence of the spheres is clear; he regarded them as mathematical fictions. In choosing combinations of devices, he regarded all combinations as equivalent if they gave the same results. He saw no significance in the differences between equivalent combinations. In fact, he calculated the positions of the planets twice for each one. The first calculation was based on the eccentric and the second on the epicycle (or equant, as necessary). He took care to point out the equivalence of the two results. His interest was solely, as he said, in "saving the appearances". The physical realities behind the motions were no concern of his.

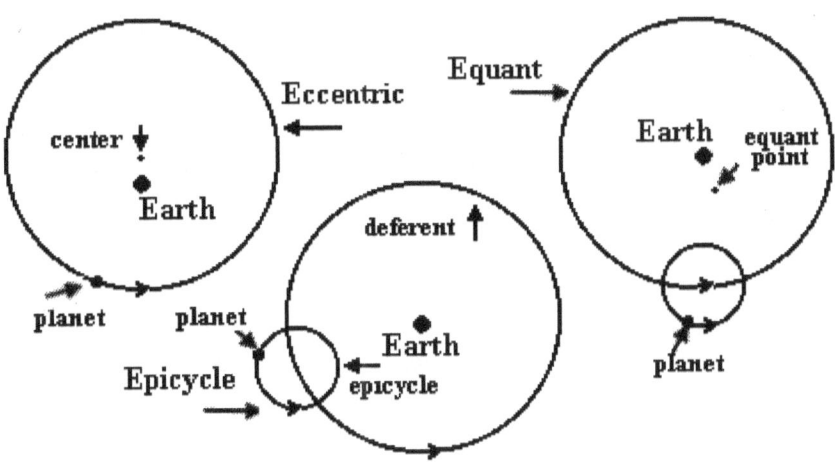

Figure 11. The three devices of the Ptolemaic system.

Dissenting from Aristotle and Plato as to the order of the planetary orbits, Ptolemy said:

> The foremost mathematicians agree that all these (planetary) spheres are nearer the earth than the sphere of the fixed stars, and farther from the earth than that of the Moon; that the three-of which Saturn's is the largest, Jupiter's next earthward, and Mar's below that-are all farther from the earth than the others and that of the sun. On the other hand, the spheres of Venus and Mercury are placed by the earlier mathematicians below the sun's, but by some of the later ones above the sun's because of their never having seen the sun eclipsed by them.... Since there is no other way of getting at this because of the absence of...(known) linear distances...the order of the earlier mathematicians seems the more trustworthy, using the sun as a natural dividing line between those planets which can be any angular distance from the sun and those that cannot but which always move near it.[xi]

Ptolemy calculated distances within his system in multiples of the distance between the Earth and the Moon or, sometimes, in units of Earth radii. As to relative sizes of these distances, we have his remarks that:

> Now, that the earth has sensibly the ratio of a point to its distance from the sphere of the so-called fixed stars gets great support from the fact that in all parts of the earth the sizes and angular distances of the stars at the same times appear everywhere equal and alike for the observations of the same stars in the different latitudes are not found to differ in the least...(also)...everywhere the planes drawn through the eyes, which we call horizons, always exactly cut in half the whole sphere of the heavens. [xii]

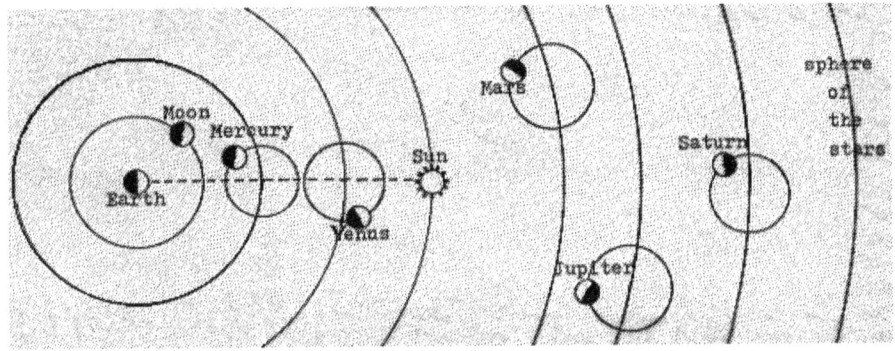

Figure 12. A Simplified Sketch of the Ptolemaic System.

A critical feature of his system was the necessity of placing Venus and Mercury on epicycles with centers on a line connecting the Sun and the Earth (Figure 12). This was necessary to keep the two planets in the vicinity of the Sun where they are always observed to be. It proved to be the Achilles' heel of the system when Galileo later discovered the full phase of Venus.

The system was extremely flexible. The numbers and combinations of devices were unrestricted so devices could be added or removed as needed without threatening the overall structure of the theory. Thus, an enormous range of orbits could be replicated (Figure 13). Ironically, Ptolemy seemed unaware that the epicycle generates the eccentric. He actually calculated pairs of orbits for each planet one based on epicycles and the other on the eccentric.

Most importantly, the epicycle structure explained a peculiar feature of planetary motion known as retrograde motion. Retrograde motion is an apparent deviation of a planet from its normal leftward motion against the stars night by night. Roughly once a year, a planet thus seems to move backward for a few nights before continuing its usual motion. Epicyclic motion easily generates a looping motion (Figure 13, bottom right) which, viewed from the Earth must appear as retrograde motion. In fact, epicyclic motion generates retrograde motion too easily and great care must be taken to avoid predicting too frequent retrograde motion of the planet.

By the criteria of longevity and ingenuity, the Ptolemaic system must be rated among the great intellectual achievements of man. Like all its predecessors it was geocentric with the Earth central to a vast universe bounded by the sphere of the fixed (unmoving) stars. It also shared with all earlier theories a lack of interest in physical causes.

The Interlude

For centuries after Ptolemy, little or nothing was done in cosmology. The West was not even aware of the Ptolemaic system until contact with the Arabs brought the *Almagest* to western attention. By the time of Dante (1265-1321) the Ptolemaic cosmology was so firmly entrenched in western thought that Dante could build his Divine Comedy around it.

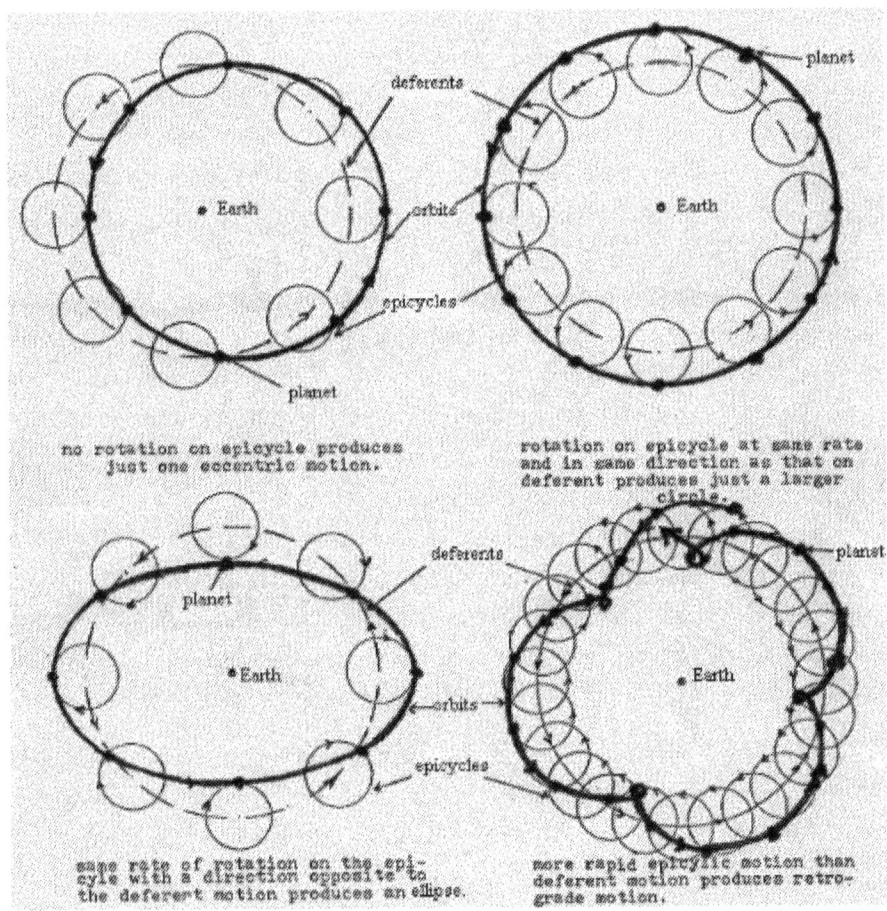

Figure 13. Some of the large variety of orbits possible with simple epicyclic motion.

Around that time, Oresme considered the possibility suggested first by Heraclides that the Earth rotates. He pointed out that a man in the Heavens would see the Earth as rotating because motion is relative. Hence, the fact that we see the Heavens rotate does not prove that they rotate. He also realized that the rotation of the Earth would not cause great winds if the atmosphere moved with the Earth and he noted that enormous speeds were required of a rotating sphere of stars while a rotating Earth entailed much more reasonable speeds. He even argued that Scriptural remarks (e.g., Ec.1:5) are not evidence for a stationary Earth. He did decide in favor of an Earth at rest but he was careful to note that that opinion was more a matter of faith than a demonstrated conclusion.

The rotation of the Earth became a common topic of scholastic consideration from that time on. The almost inevitable conclusion was that the rotation was surely impossible. Nonetheless, the frequency with which the question was raised suggests that the idea was not really inherently implausible to the schoolmen.

Nicolas of Cusa reviewed Epicurus' argument that the universe is infinite and has no boundary. He concluded that there is no physical center of the universe and happily echoed Empedocles who wrote "God is an infinite sphere whose center is everywhere and circumference nowhere". With no physical center to the universe, Cardinal Nicolas felt free to accept the idea that the Earth rotates about its axis. Nicolas was a nominalist, not a materialist like Epicurus. To him, only God could be infinite so the universe, he decided, must be indeterminate. He did retain the Epicurean idea of the multiplicity of worlds.

Throughout this period, there was a lively interest in the Antipodes. These were people who purportedly live on the opposite side of the Earth from "us" so their feet are above their heads (hence, Antipodes). Lucretius thought they would fall off the Earth and a number of later writers followed his lead, ridiculing the idea of a spherical Earth without proposing an alternative model. Boniface, Archbishop of Mentz in the eighth century complained to Pope Zachary that Virgil, Bishop of Salzburg believed the Antipodes exist. Boniface was alarmed because people in such a place were (he supposed) out of reach of missionaries and salvation. The Pope apparently ignored the complaint. Shortly before the voyage of Columbus, Tostatus condemned belief in a spherical Earth as "unsafe".

One sixth century writer, the geographer and explorer Cosmas Indicopleustes, actually proposed a flat Earth like the floor of a room with walls. The heavens formed a ceiling like vault over the room and the stars and planets circled below the vault. An enormous mountain in the north caused night when the Sun circled behind it. A moment of reflection shows the east to west motions of heavenly bodies cannot be reproduced by this scheme.

The Copernican Revolution

Nicolas Copernicus (1473-1543) studied medicine and canon law at the University of Padua but took his degree from the University of Ferrara, though he never studied there. The reason seems to have been the price; the University of Ferrara charged much lower graduation fees! He must also have studied astronomy for, though canon law got him his income and his knowledge of medicine made his reputation among his contemporaries, it was his devotion to systematizing astronomy that has made him historically important. Evidently his dissatisfaction with the Ptolemaic system began with his first encounter. He found the equant unesthetical and its non-uniformity seemed inconsistent with the effort to see the heavenly motions as perfect. Ptolemy's inconsistent choosing of different combinations of devices to solve similar problems also offended Copernicus.

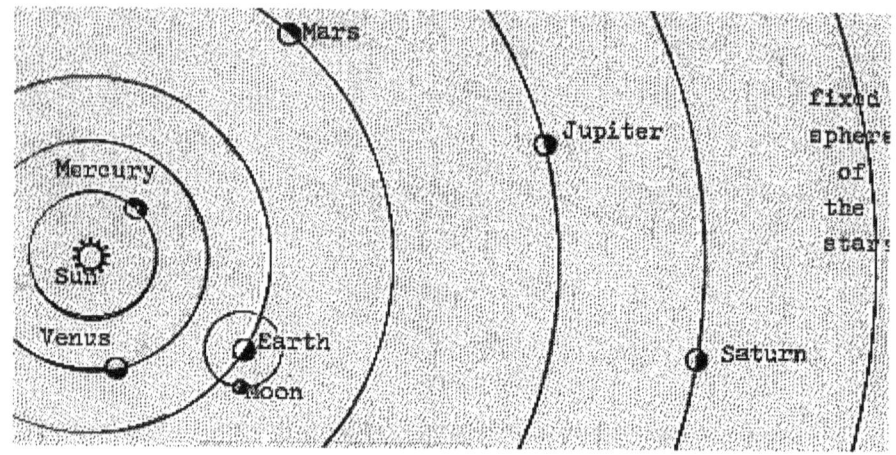

Figure 14. A simplified sketch of the Copernican system.

Trying to improve the system, Copernicus found it necessary eventually to alter the basic features of the system. He was able to eliminate the equants at the price of removing the Earth from the center of the universe, though it remained the center of gravity. All the spheres were made to revolve around the Sun but there was no one center of all the spheres (Figure 14). The Earth rotated on its axis and revolved around the Sun, accounting for the diurnal motions of the heavens and many of the other planetary peculiarities as well. Copernicus particularly remarked that the stars were very distant compared with the distance between the Earth and the Sun. This was an effort to pre-empt his critics because Copernicus knew that no parallax could be found in stellar positions from one side of the Earth's orbit to the other (no annual parallax). Nevertheless, his critics preferred to believe the lack of parallax was because the Copernican system was false. The enormous size of the Copernican universe was unacceptable.

The new system did provide a neat order unlike anything available in the Ptolemaic system. The nearer a planet is to the Sun in the Copernican system, the faster it moves. Allied to this advantage of the Copernican scheme was the causal explanation it provided of retrograde motion, an achievement much superior to the mere replication of the motions characteristic of the Ptolemaic system (Figure 15). Retrograde motion is an optical illusion occurring as the Earth catches up with and then passes another planet on its orbit.

Copernicus also had a natural explanation of the proximity of Venus and Mercury to the Sun. As planets interior to the Earth's orbit, they necessarily always appear near the Sun. However, the number of spheres in both systems is about the same. Copernicus defended heliocentrism, saying that it was appropriate that the light giver, the symbol of God himself, should be central.

Copernicus made another suggestion possibly more astonishing than even his heliocentrism. He accounted for the spherical shape of the Sun, Moon and planets by assuming the gravitational argument by which Aristotle showed the Earth was spherical could properly apply to the heavenly bodies also! The implication was that more than one center of gravity might exist! The Aristotelians thought this was very dangerous stuff as, indeed, for them, it was.

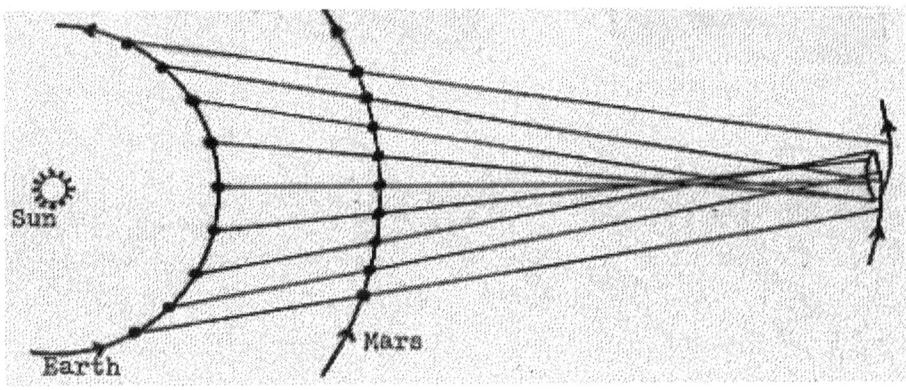

Figure 15. The Copernican explanation of retrograde motion. The position of Mars seen against the fixed stars at the right seems to go "backwards" as Earth catches and then passes Mars.

His contemporaries were unimpressed. For them, the difficulties were insuperable. The system defied Aristotelian physics and metaphysics, Scriptural passages and the views of the church fathers and common sense. The minor problems of Aristotelian physics and of Ptolemaic cosmology were not nearly important enough to cause anyone to abandon the system. Additionally, there was no new physics to replace Aristotelian physics.

Writing in English in 1556, Thomas Digges advocated the Copernican system. He went on to make the astonishing suggestion that the stars were not all at about the same distance from Earth but were scattered throughout the cosmos at various distances from Earth. Aristotelians and Copernicans alike were appalled.

Tycho Brahe, toward the end of the 16th century, weighed in with an alternate system (Figure 16). He revived and expanded ancient ideas, making all the planets (except the Moon) orbit the Sun. The Sun then orbited the Earth. This system, which was probably not original with Tycho, did retain a general geocentrism and explicitly had the Sun orbiting the Earth in supposed agreement with Scriptures. Other systems were suggested so that by 1600, there were more than five systems known to scholars.

Johannes Kepler, Tycho's assistant, promised the dying Brahe that he would work to support the Tychonic system. Of course, he did no such thing but improved the heliocentric system by discovering three precise laws of planetary motion about the Sun. They are: a) The planets travel on elliptical paths with the Sun at one focus. So nearly circular are these ellipses that in all cases except Pluto (which was unknown at that time) the second focus is also inside the Sun. That is why the idea of circularity lasted so long. It was almost right! b) The planets sweep out equal areas in equal times (Figure 17). This implies that the closer a planet is to the Sun, the faster it moves. c) The square of the period of the orbit is directly proportional to the cube of the radius of the orbit (semi-major axis, for an elliptical orbit). In equation form the law is $T^2 = ka^3$. This law also implies that the closer planets move faster but it also provides a way of comparing the planets which was not available from the 2nd law.

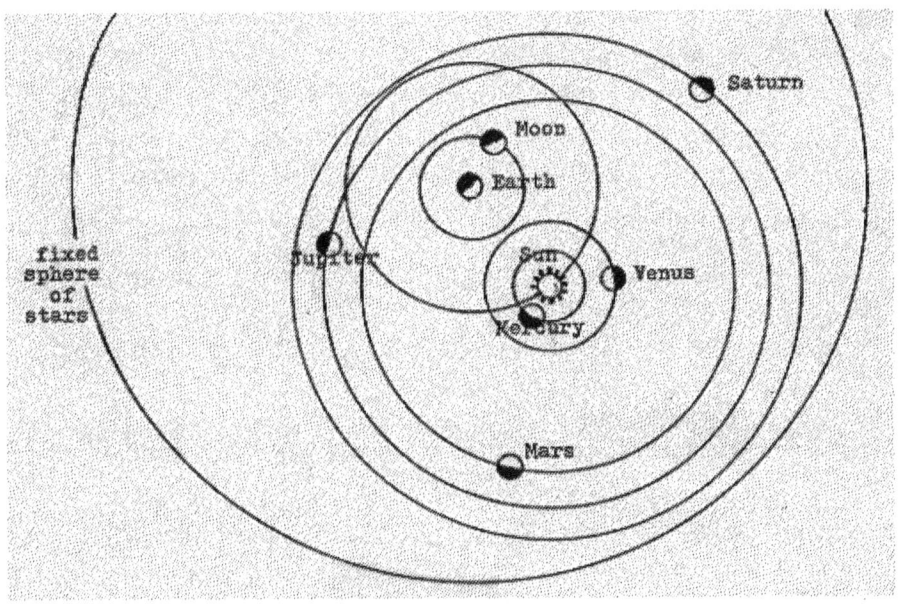

Figure 16. Tycho Brahe's compromise system. The planets go around the Sun which goes around the Earth.

Kepler made another contribution which is harder to evaluate. William Gilbert, the personal physician of Queen Elizabeth, published a very important little book, ***De Magnete***, in 1600. In it he demonstrated convincingly that the Earth is a magnet. Much impressed, Kepler guessed that the Earth was held in its orbit by a magnetic force emanating from the Sun. Furthermore, he explicitly saw gravity as a mutual attraction, saying that "...the Earth draws a stone much more than the stone draws the Earth..." This was the first time anyone had thought the stone could draw the Earth. Further, Kepler suggested that two stones would gravitationally attract each other and move together. The amount of movement each would make would be controlled by its "bulk"! Please note: Kepler did not realize that it is gravity that holds the Earth in orbit about the Sun. Also note he saw the attractions as mutual, just as each magnet attracts the other. In all this he prefigured Newton.

Before reading ***De magnete***, Kepler evidently thought the planets were moved by vortices of the matter between the planets. The rotation of the Sun caused these vortices. René Descartes developed this vortex idea more fully.

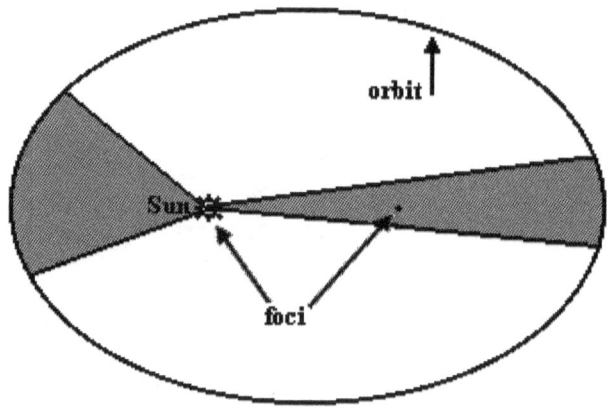

Figure 17. Kepler's Second Law of Planetary Motion. The two areas in gray are equal so the distance along the orbit and planetary speed are greater for the area closer to the Sun.

The telescope first appeared around 1608 and by 1610 Galileo published a book, ***The Starry Messenger***, in which he described an amazing list of discoveries he had made by turning this device on the heavens. He described mountains on the Moon and even estimated their height at about 4 miles (a reasonable value). He timed the motion of sunspots and was able to give a rotation rate for the Sun as well as showing that the spots were indeed features of the surface of the Sun and not, as some thought, planets very close to the Sun. (The spots had to be surface features because Galileo saw them flatten as they reached the edge of the Sun). Both these discoveries implied imperfections in the heavens which the Aristotelians of the day regarded as impossible.

Looking at the Milky Way, he was the first to demonstrate that it is composed of almost innumerable stars. Democritus had guessed this fact 2000 years before, possibly seeing atomistic implications in it. Galileo also discovered the four largest moons of Jupiter. Strangely, the Aristotelians regarded these discoveries as changes in the heavens and violations of the immutability of the heavens! Galileo himself was very impressed with the moons because he saw in them a proof of the Copernican system. The moons circling Jupiter seemed a miniature model of the larger solar system. His contemporaries did not see it that way.

Galileo's most important discovery with the telescope was that Venus has phases like those of the Moon (Figure 18). In particular, the full phase when Venus is almost behind the Sun from Earth simply could not happen in the Ptolemaic system where the planet is never beyond the Sun as viewed from Earth. Hence, this one observation operated as a clear counter-example to the Ptolemaic system which then went into a permanent decline. The major Jesuit astronomers quietly shifted to the Tychonic system. In their view, the Copernican system was too problematic. Galileo was very put out with them, thinking he had proven the Copernican system.

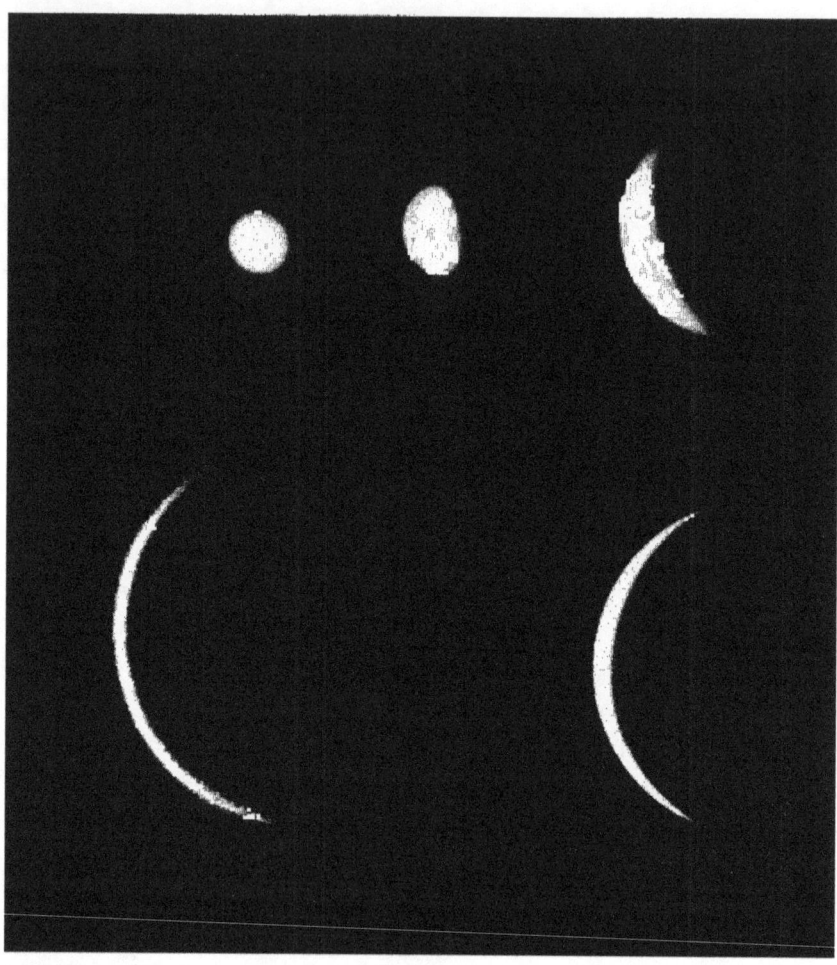

Figure 18. A composite picture of the phase of Venus. Note the small, full phase that is impossible in the Ptolemaic system.

Isaac Newton took the final, critical step to complete the Copernican Revolution. Copernicus and Kepler provided the right model but the model made no sense in terms of Aristotelian physics. Newton provided the new physics, producing at the same time a causal explanation of planetary motions. The cause was gravitation. This was not the old gravity of the Greeks but a new force, acting over the great cosmic distances between celestial bodies and acting on the celestial bodies as Aristotle never could have thought possible.

The story of Newton and the apple is well known. Two features of this experience are not generally recognized. The usual version has Newton "discovering" gravity although the Greeks knew gravity as the natural tendency to free fall. What Newton discovered was much more subtle. Thinking of the apple as a wind-blown projectile helps in understanding just what speculations the event inspired in Newton. Newton must have wondered what would happen as the wind increased in strength. Since the apple falls further from the tree as the wind increases, Newton realized that a very strong wind would take the apple around the world on a closed path. But such a path already occurs for the Moon so he was inspired to further speculate on the possibility that the Earth's gravity extended to the Moon and held it in its orbit. He claimed to have then made the calculation and found it to agree "pretty nearly" with observation.

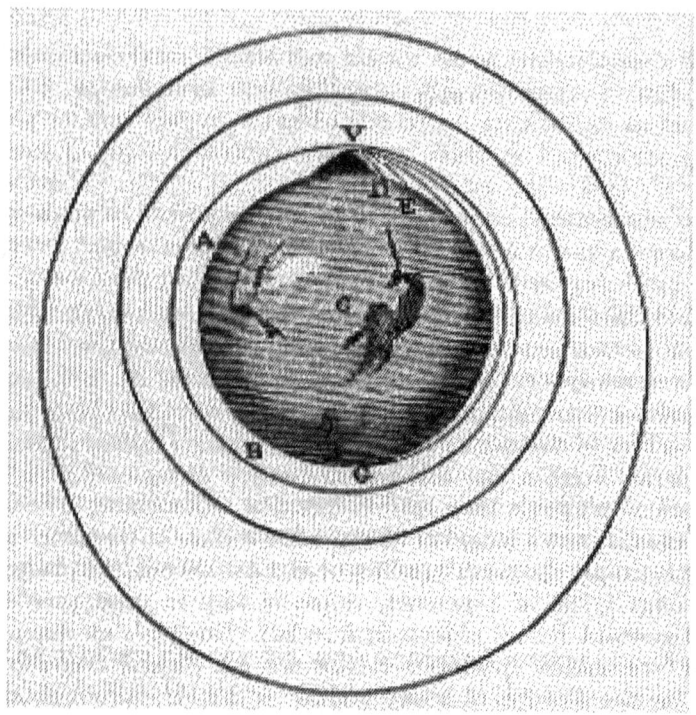

Figure 19. Newton's thought experiment. The sketch shows that projectile motion with sufficient horizontal speed becomes orbital motion, the Newtonian Synthesis.

Thus Newton realized orbital motion is an extension of projectile motion (Figure 19). Now, Galileo had shown projectile motion to be a combination of vertical free fall with a constant speed motion horizontally. Thus, what for Aristotle had been three distinct motions were transformed into three cases of the same basic type of motion. We call this intellectual achievement the Newtonian Synthesis.

However, before he could make progress with this idea, he needed to have a mathematical expression of the strength of the gravitational force at various distances from the body originating it. Is the Earth's gravity as great at the Moon as here at the surface? He was sure it was not but how could one find out exactly what it is and how could the answer be used? A force that changes with distance acting on a body moving at continuously changing distances and speeds lay outside the mathematics of the day. Newton was forced to invent a new mathematics, the calculus, to do the calculations. Assuming an inverse square variation with distance, he was able to deduce Kepler's laws of motion. Thus, the form of the law was confirmed. Newton further surmised that the source of the gravitational force was the mass of the bodies. Taking all these considerations into account, the law of gravitation had to be:

$$F = Gm_1m_2/r^2$$

In the law, the masses m_1 and m_2 are those of the two bodies involved in the gravitational interaction, and r is the distance between their centers. G is the universal gravitational constant measured later by Henry Cavendish and now known to be 6.67×10^{-11} Nm^2/kg^2. The law is called the Law of Universal Gravitation because it implies that any object having mass also is a source or center of gravity.

The law produced tremendous successes but it has not been uncontroversial. Universal gravitation introduced a problem that earlier systems had not had. If every object attracts every other object, there should be a tendency for all bodies in the universe to aggregate into a clump. The problem was anticipated by Lucretius when he suggested that "matter would already have accumulated by its own weight at the bottom".

Digges' suggestion that the stars were not at a single distance from the earth but were scattered far apart helped to a degree if the distances were assumed very great and the universe were assumed quite young, because then the clumping might not yet have had time to occur. Lucretius avoided this aggregation by denying boundaries or a center to the universe. Newton, too, used the idea of an infinite universe to dodge the problem. He argued that in an infinite universe with an infinite number of stars randomly placed, there would be no center of mass for the stars to fall toward. He said:

It seems to me, that if the matter of our sun and planets, and all the matter of the universe, were evenly scattered throughout the heavens, and every particle had an innate gravity towards all the rest, and the whole space throughout which this matter was scattered, was finite, the matter on the outside of this space would by its gravity tend towards all the matter on the inside, and by consequence fall down into the middle of the whole space, and there compose one great spherical mass. But if the matter were evenly disposed throughout an infinite space, it could never convene into one mass; but some of it would convene into one mass and some into another, so as to make an infinite number of great masses, scattered great distances from one another throughout all that infinite space. And thus might the sun and fixed stars be formed.[xiii]

Nevertheless, stability of the universe is a problem once universal gravitation is admitted.

In Newton's own day, the Cartesians angrily complained that Newton had reintroduced the animism and superstition Descartes had banished from the world. Newton's law implied gravity acts instantly across a vacuum. This "action at a distance" undercut the mechanical Cartesian idea that a medium was necessary to convey force from one place to another. Also, the transport of force would require time to travel in the medium. Worse yet, the use of the word "force" seemed to explicitly invoke the old, mystical, Stoic ideas of the world. Newton said he was sorry, the theory just did not seem to need Descartes' medium or time and he murmured that he would "frame no hypotheses" ("hypotheses non fingo") as to why this was so. To make matters worse for the Cartesians, Newton then proved unworkable Descartes' idea that a great vortex created by the Sun in the interplanetary medium carried the planets around on their orbits. He showed that the actual speeds of the planets are inconsistent with the speeds such a vortex would produce.

In the twentieth century, Einstein's General Theory of Relativity largely supplanted Newton's theory of gravity and gravity is now viewed as a force that is transferred at the speed of light possibly by a graviton particle. The Cartesians would be pleased.

From the time of the publication of Newton's work and throughout the rest of Newton's life, Copernican cosmology and Newtonian physics rapidly replaced Ptolemaic cosmology and Aristotelian physics. The new views could explain successfully all the details explained by Aristotelian ideas. In addition, the new views explained new details that the old ideas could not accommodate. The parabolic path of projectiles, the elliptical orbits of the planets, the full phase of Venus and the distance of comets and super novae from the Earth all made sense within the new physics and cosmology and had not made sense from an Aristotelian point of view. The revolution was complete; Newton reigned supreme.

The Big Bang

The eighteenth and nineteenth centuries were a period of plateau in cosmology. Though a great deal of work was done and many discoveries were made, it was basically a time of working out the details of Newtonian mechanics. As the twentieth century began to probe the structure of the universe beyond our Solar System and its immediate environs, the situation began to change; in particular, the Big Bang theory provided a startling, new vision of the structure of the cosmos.

The beginnings of the Big Bang theory of the universe go back to the work of Vesto Melvin Slipher (1875-1969), an American astronomer. Galaxies, which had been earlier called nebulae, were beginning to be suspected to be vast aggregations of stars. In 1913, hoping to show that galaxies rotate, Slipher measured wavelengths of light coming from the edges of a number of nearby galaxies. His thought was that one edge of a rotating galaxy should rotate towards us and the opposite edge should rotate away. The difference in rotation should appear as a difference in the wavelength of the light from one edge to the other.

He never was able to detect the difference but he did find that all the galaxies radiate red shifted light. That is, light coming from these objects has a longer than expected wavelength indicating the light source is receding. Accordingly, he reported to the American Astronomical Society that these galaxies were moving away from the earth. Slipher continued this work for another decade. By 1925 he had measured a recessional speed for 42 galaxies.

Another part of the story is the Theory of General Relativity which began around 1916 when Albert Einstein (1879-1955) wondered why mass should be a source of gravitational force. The result was a new view where a mass actually deforms or curves the space around it and the curvature of that space moves other masses toward the first mass. Solving the equations derived from this idea, Einstein found that solutions of his equations implying the universe should be expanding. It was the old stability problem back again.

At that point Einstein made what he later called the "greatest blunder" of his life. Unable to see any physical sense in a expanding universe solution (after all, it doesn't look like it's expanding-any more than the Earth seems to move), Einstein introduced a "cosmological constant" into his equations. Although Plato's circles and geocentrism were long gone, belief in Aristotle's eternal and static universe was still very deeply held by Einstein and most physical scientists of the time. The sole purpose of the constant was to cancel out the expansion and turn the expanding universe solutions into static universe solutions.

It was not long, however, before Alexander Friedmann (1888-1925) in Russia discovered another *expanding* universe type solution. Georges LeMaitre (1894-1966), a Belgian priest who had studied physics and astronomy under Sir Arthur Eddington (1882-1944), having independently discovered the same solution as Friedmann, proposed an expanding universe model which began with a primordial Big Bang. At first, Einstein resisted vigorously, thinking the results mistaken but eventually he realized they were right.

About the time Slipher stopped clocking galaxies, Edwin Hubble (1889-1953), who had been in the audience the day Slipher related his startling news of receding galaxies, began a new research effort. It had been suspected for some time that the "nebulae" which, through a telescope, are fuzzy and indistinct compared with stars are really vast and enormously remote collections of stars like our Milky Way galaxy. Using the new 100 inch telescope at the Mount Wilson Observatory, Hubble set out to decide the issue. His photographs showed without doubt that the spiral nebulae are collections of stars.

Hubble concentrated next on the brightness and rate of brightness fluctuation of individual Cepheid variable stars within a group of galaxies. Work done in the previous decade by Henrietta Leavitt (1868-1921) and Harlow Shapley (1885-1972) had shown the existence of a relationship between the true brightness of Cepheid variable stars and their period of variation. Measuring the period of the variations of variable stars in the galaxies he was studying gave Hubble the intrinsic brightness of the variables. Comparing it with the measured brightness enabled him to calculate the distance of the variable and its galaxy.

Hubble then persuaded Milton Humason (1891-1972) to measure the red shift of the light from these galaxies. Comparing the two sets of data, Hubble found a direct relationship with the more distant galaxies having the largest red shifts. The shift is due to the motion of the galaxy and the fact that it is a shift toward longer wavelengths implies that the motion is away from us.

The most obvious interpretation of this that made any sense to Hubble was that all the galaxies are receding from us and the farther away they are, the faster they are moving, the pattern of LeMaitre's explosion. The implications were that the universe had begun with an explosion whose effects are still apparent in the gross motions of objects in the universe. The universe began with a Big Bang.

His fellow astronomers did not greet the Hubble expansion with whole-hearted enthusiasm but there never has been a serious alternative interpretation of his red shift and Cepheid variable data. On the other hand, the Big Bang was greeted with definite hostility. Expanding universe solutions of the equations of General Relativity were merely suggestions of what is possible. The Big Bang Theory was too much like actuality and seemed to imply the universe had a beginning. The static universe bias had to be dealt with explicitly. The avowed atheist, Fred Hoyle (1915-), declared that the Big Bang was entirely too much like a theistic creation. Like the Cartesians before him, he felt science was letting in by the back door something it had banished from the universe by the front door. Challenged by the new ideas, Hoyle collaborated with Thomas Gold (1920-) and Hermann Bondi (1919-) to invent an alternate explanation called the "Steady State Universe" where the Hubble expansion was allowed but a quasi-static universe was retained by the expedient of having atoms of hydrogen periodically pop into existence to maintain a constant density of matter in the universe. Why he felt this continuous creation *ex nihilo* preferable to a single creation event is not obvious. He argued that the continuous creation of hydrogen atoms was at least testable in principle.

In the 1950's, George Gamow (1904-1968) and his students examined the Big Bang idea and noticed the original fireball of the Big Bang should have produced bright light that should still be around today though it would no longer be visible since its wavelength should increase with the increase in size of the universe. As the wavelength of visible light increases, the light changes from visible light to infrared light to microwave radiation. They predicted this fossil of the Big Bang should now be microwave radiation coming from every direction in the sky with the characteristics of light from an object at about 5 degrees above absolute zero. They also found the Big Bang nicely explained the ratio of helium to hydrogen in the universe.

There the matter stood for a number of years until, in 1965, Arno Penzias (1933-) and Robert Wilson (1936-) made a lucky discovery. They were working at Bell Labs on microwave transmitter/receiver horns for astronomy and were having problems getting rid of a persistent noise in the reception of their sets. Eventually, talking with colleagues, they discovered that a search was on for a radiation that George Gamow and others had predicted might be left over from the original fireball of the Big Bang.

Investigation of the problem revealed that the sizzle in their reception was indeed the fossil radiation of the Big Bang so heavily red shifted by the Hubble expansion that it looked like radiation emitted by a body at 3 degrees above absolute zero (3 K). Hence, it is known as the 3 K radiation (although recent satellite data shows very beautifully that the radiation is indeed blackbody radiation at about 2.7 K). The Steady State Theory proved unable to account for the 3 K radiation and leaving the Big Bang Theory as the only available theory of the formation of the universe. Modifications to the theory have already appeared but the basic ideas are little altered.

The stability question is resolved in the standard Big Bang model. The gravitational attraction is present and has been present since the beginning of the universe, about 15 billion years. However, even in that long time, the attraction has not caused the stars to clump together and "be lying together in a heap throughout eternity" as Lucretius suggested would happen. The great speeds of the Big Bang are being slowed by gravity but they have not yet been reduced to zero. Like a ball thrown into the air, the pieces of our universe are moving away but slowing down. Will they ever stop, like the thrown ball, and return together? The available data are inconclusive but, at the moment, most likely the Hubble expansion will go on forever.

Further Reading

Constructing the Universe, David Layzer, Scientific American Books, Inc., 1984.

Kepler's Geometrical Cosmology, J.V. Field, Univ. of Chicago Press, 1988.

The Discoverers, Daniel J. Boorstin, Random House, 1983.

Chapter IV

Motion and Time

The Classical Period

It is arguably fair to say that much of Greek metaphysical thought was an attempt to answer the question, "What is constant in a world of change?" To an extent the question was culturally conditioned. Economic and political instabilities encouraged such questions if they did not obviously trigger them. However, the question is natural. We all want to know what will change and cannot be counted on, and what can be trusted to remain the same. Since motion is one type of change, motion became one of the phenomena of which the Greeks sought explanations.

Most of the Ionians responded by positing some basic element as the unchanging stuff of the world. In effect, the invention of elements answered the question with the reply, "The elements never change; change is an interchange or motion of the changeless elements." Water for Thales, air for Anaximenes, and the "boundless" for Anaximander; the basic stuff changed through processes such as condensation or melting. The processes were the causes of motion. Thus began a centuries-long dialogue among Greek thinkers, a series of responses and counter-responses.

An Ionian with a different, disturbing thought was Heraclitus of Ephesus. "*The riddler*", as he was called, insisted that change is all there is. Change is caused by conflict, a war of opposites. The material embodiment of this strife is fire.

Elea, a Greek colony in southern Italy, produced a number of important thinkers. One of them, Parmenides (b.515 BC), may have visited Athens and met Socrates. He placed stress on the law of contradiction. If, as Heraclitus claimed, opposites are at war, what is the opposite of being? Non-being is the only possible answer but to say non-being exists is a contradiction and must be nonsense. Thus, no opposite of being exists and no war can occur. Nothing really changes, then, because all things already have being. Change can be only an illusion.

Wisely, Parmenides died and left to his student, Zeno (b.495 BC), the task of elaborating and defending these ideas. This Zeno did with a series of paradoxes aimed at showing the contradictions inherent in any other view of motion and change. The most famous is that of the race between Achilles and the tortoise. Retelling it in modern terms we might assume Achilles is ten times as fast as the tortoise and the tortoise has a 100 m head start. When Achilles has covered 100 m, the tortoise is 10 m ahead. When Achilles has run another 10 m, the tortoise is 1 m ahead and so on. At the end of every such time interval, the tortoise will still be ahead! The conclusion is that Achilles can never catch the tortoise, though, of course, we know he can.

The error in Zeno's discussion was not obvious to ancient hearers anymore than it is to a general audience today. He erroneously assumed that the sum of an infinite series (of time intervals) is always infinite regardless of the size of the terms. His fundamental conclusion is that Achilles will need an infinite amount of time to catch the tortoise.

The sum of an infinite number of steps (an "infinite series") is not necessarily equal to infinity. Some series do sum to infinity but many do not.

The series Zeno had in mind is not infinite and clearly has a finite sum as we can see if we add terms. The total time taken by Achilles to travel 100 m, then 10 m, then 1 m and so on is not infinite but actually not long at all. He will catch the tortoise after running 111.1 m. The time required will not be infinite, as Zeno assumed, but will simply be 111.1 m divided by Achilles' speed. Our intuition that he will quickly catch the slow tortoise is correct. But Zeno's contemporaries did not understand the root of the paradox and it troubled them inordinately. In particular, it made Greek thinkers suspicious of anything having to do with infinity or an infinite number of steps.

Anaxagorus, followed by Leucippus and Democritus, gave an atomistic response to Zeno. The error in Zeno's paradox was, they thought, the assumption that distance can be indefinitely subdivided. Eventually, the irreducible atoms will prevent the endless dividing Zeno had in mind. So the analysis was erroneous. Motion, for the atomists, was the normal state of matter, needing no explanation.

Plato's response to Zeno was to let change reside in the world of appearances while the world of forms was changeless and timeless. Thus, Plato agreed that change was an illusion in the sense that all appearance is illusion. Time was the movement of the heavens, a moving image of the form or principle of soul. Since the heavenly motions were mensurable, Plato had some sense of quantifying motion. However, there is nothing in Plato that qualifies as a law of motion unless we count the view that heavenly motions are combinations of circles.

In the teleological universe of Aristotle, changes were not illusions but either followed a purpose (natural changes) or were caused by an external agent (violent changes). The question of whether or not an external agent was involved was answered in a common sense manner. If you can see the agent or its effects, it is acting. It not, no agent is acting. On Earth, natural motions were upward for the elements fire and air and down for earth and water. Any other motion of things made of these elements was violent. In the heavens where all bodies were pure aether and no agents existed, the natural motion was circular and no violent motion occurred (the doctrine of the immutability of the heavens).

Another factor in motion was the resistance of the medium. As resistance decreased, the speed of a body increased (Aristotle's law of proportional effects). Since an infinite speed is inconceivable, a vacuum is impossible. As Aristotle said, "Nature abhors a vacuum." Thus, the Aristotelian universe was a plenum, space completely filled with matter without voids or empty spaces.

In summary, Aristotelian physics had three major laws. The *first* was the law (or doctrine) of the immutability of the heavens that the only changes in the heavens were circular motions and possibly their associated changes of apparent brightness as a planet approached or receded from Earth. *Secondly*, the law of violent motion was that violent motion ceases when the impelling force ceases. *Lastly*, the law of proportional effects was that speed of objects in a medium is inversely proportional to the density of the medium. If the medium density is doubled, the speed of an object moving in the medium must be halved. Aristotle also thought that a heavy body falls faster than a light body. Latter followers treated this belief as an indubitable law.

A major problem arose with respect to projectile motion. Since violent motion ceases with the cessation of its causative force, how does a projectile continue its motion? Aristotle cautiously suggested the medium, displaced at the front, rushes around to the back in order not to allow a vacuum. In so doing, it pushes from behind on the projectile, thus continuing the motive force. This idea of "*antiperistasis*" was the point from which Aristotelian physics later began to unravel.

Additionally, the full universe, lacking a vacuum anywhere, created a cosmological problem for Aristotle because friction between the heavenly bodies should occur due to the celestial medium. In order to eliminate this friction, Aristotle had to add "unrolling" spheres to those already proposed by Eudoxus. An unrolling sphere had the unenviable, if not impossible, assignment of communicating the motion of the rotating stars to a planet without communicating any frictional effects of the intervening medium.

Lastly, while time is connected to movement and change, Aristotle distinguished it from them. Time was a continuum best measured by uniform motions. The heavenly motions, as the most uniform known motions, were the best choices for keeping time. Counting time by the solar movements was thus seen as the best system.

Strato, who became head of Aristotle's Lyceum in 286 BC, conceived acceleration as an increase in speed and argued that water streaming from a roof must be accelerated. Noting that it is first a stream but then breaks into droplets, he argued that acceleration stretches the stream out until it is too thin to be a stream. Acceleration stretches the stream out because different sections of the stream have different speeds and the stretching leads to breaking up. He also argued that there is no natural tendency to rise. The rising of air and fire is due to displacement as heavier air masses fall.

The Interlude

John Philoponus in the early sixth century AD took issue with a number of Aristotle's ideas of motion. Aristotle had recognized exceptions to his rule of proportional effects of a medium. For example, he noted that if twenty men could move a boat twenty feet in a minute, one man would not necessarily move it one foot in a minute. He might be unable to move it at all! Philoponus added falling bodies to the exceptions by pointing out that a mass twice as heavy as another simply does not fall in half the time but in almost the same time.

As to projectiles, Philoponus thought it made more sense to attribute the continuing force to the projectile itself rather than to the medium. Since blowing on a rock will not move it, how can air keep a rock in motion? He speculated that the original motive force created an inherent, internal force called "impetus" which continued the motion of the projectile but declined throughout the motion so that the projectile eventually landed. Not to be outdone, Philoponus' contemporary and rival for the title of last classical scholar, Simplicius of Cilicia, suggested the air and the projectile might alternate in providing the impelling force to continue projectile motion!

Arab thinkers added to these counter-Aristotelian ideas. Abu Ali al-Husayn ibn Abd Allah ibn Sina (980-1037), called Avicenna in the West, thought the impressed force produced a "desire" in a projectile which was decreased by the resistance of the medium so that, in a vacuum, the motion would be constant. Furthermore, he saw gravity as a natural desire similar to the desire of the projectile so that a free falling object, like a projectile, would move out of a desire to move. This view, therefore, blurred the absolute Aristotelian distinction between free fall and projectile motion. Avicenna was known as the Prince of Physicians and his *Canon of Medicine* was used as a medical text into the seventeenth century. Later, Avempace (ibn Bajja, complete name unknown)) explicitly regarded the medium as subtracting motion from the projectile.

The ideas of both these men reached the West in the commentaries of Abu al-Walid Muhammad ibn Ahmad ibn Muhammad ibn Rushd (1120-1198). Averroës, as he was called in Europe (exclusive of Spain where he was born and lived), took issue with Avempace who thought the stars should move with infinite speed because of the lack of resistance by the celestial medium. Aristotle might have objected that there is a medium and there is resistance so there can be no infinite speed. Averroes, more of a Platonist, grouped the celestial bodies with animals because to each he ascribed an intelligence. The intelligence supplied the motive force and the body supplied the resistance. Hence, though agreeing with Aristotle that the stars did not move at infinite speeds, unlike Aristotle, Averroes had no need of a celestial medium.

Thomas Aquinas also thought the medium subtracted from motion but he still believed a medium was necessary to continue violent motion. He disagreed with Aristotle about a vacuum, arguing that a vacuum was no less likely than a plenum. He ascribed extension and dimensionality (the taking up of space) to the vacuum and took the important step of noting that motion must be considered with respect to a fixed frame of reference.

A considerable advance occurred when William of Ockham distinguished between *being in motion* and *being moved*. This separated the description of motion (kinematics) from the explanation of motion (dynamics). That is, if *being in motion* is not the same thing as *being moved*, then *being in motion* (which requires only description) can be studied separately from the study of *being moved* (which requires an understanding of the causes of motion). The Mertonians assimilated Ockham's distinction and found themselves free to concentrate on the description of motion without having to understand the causes of motion. They were thus enabled to make the progress we have already noted. Bradwardine came to understand that two bodies of the same density but very different weights would fall at the same speed in a vacuum.

Jean Buridan (1300-1358) revived the impetus theory of Philoponus. A student of William of Ockham and later rector of the University of Paris, Buridan is often given mistaken credit for single-handedly inventing the impetus theory. The Aristotelian concept of *antiperistasis* seemed a hopeless muddle to Buridan who cited three examples of situations where it could not possibly explain a continued motion. The projectile motion of a spear sharpened at both ends could not be continued by air rushing around from in front to push on the back because there is no back end to the spear. Secondly, a millstone set in motion continues to rotate for a long time after the agent of the motion ceases acting but there is neither front nor back to the object. Lastly, sailors on a ship gliding forward but no longer under sail experience a wind coming from the bow (front) but not from the stern as one would expect if antiperistasis were continuing the forward motion of sailors and ship.

Buridan defined impetus as the product of the speed and quantity of matter in the body. It caused the continuation of motion. Unlike Philoponus and Avicenna, Buridan viewed impetus as a permanent quantity given the moving body rather than an evanescent quality that the object gradually lost. However, impetus could be taken away from, as well as be given to, a body so its permanence was highly qualified. Motion on a curved path required curvilinear impetus and linear motion required linear impetus, thus preserving the distinction between natural and violent motions. Buridan explained the continuing motions of celestial bodies by assuming God had supplied them with their initial impetus which then continued the motions forever without the need for on-going divine intervention.

The rise of clockmaking around 1300 began to produce a more structured view of time and made possible a mechanical model of the universe as opposed to earlier animistic models. Thus, the "clockwork image" of a strictly mechanical universe became possible. We find Oresme saying,

> When God created…He impressed them [the heavens] with a certain quality of force and motion, just as He impressed terrestrial things with weight…; it is just the same as a man building a clock and leaving it to run itself. Thus God left the heavens to be moved continually…according to the order established.[xiv]

Thus, the clockwork image combined with Buridan's idea of the permanence of impetus to give birth to the modern view of a self-sufficient cosmos running on its own.

The Scientific Revolution

At times, progress in science has only been possible when careful attention has been paid to hair-splitting detail. Ockham's distinction between *being moved* and *being in motion* allowed the Mertonians to ignore causes of motion and to concentrate on describing motions accurately. A second critical distinction had to be made before more advances were possible in the study of motion. From Aristotle to the seventeenth century, discussion of motion had focused on quantity of motion and largely overlooked change of motion. Impetus was essentially a quantity of motion. The modern concept of momentum is a near equivalent. In the work of Galileo, the emphasis switched from questions of why motion continues to questions of how motion changes. However, no more than the Mertonians did Galileo tackle the question of why motion changes. Causes remained out of reach until Isaac Newton, standing as he said "on the shoulders of giants," stood high enough to grasp them. Galileo was one of Newton's giants.

It is tempting but mistaken to think that Giovanni Benedetti (1530-1590) was instrumental in affecting this change in thinking. His writings show that he thought a falling object continually increases its speed because gravitational force is continually applied to it. Benedetti seems to have been personally unpopular with Tartaglia, his teacher, and with many of his peers and this may explain why his work was largely ignored. That is unfortunate because he did some good work on falling objects. He (not Galileo) was the first to show and then state plainly that Aristotle was wrong about proportional effects because bodies made of the same material fall at the same rate regardless of weight. He also was first to note that resistance of a medium on a body moving through it depends on the surface area of the body and not on the volume or weight of the body. Lastly, he pointed out that a body released from circular motion will follow a straight line tangent to the circular motion. It seems unlikely that Benedetti influenced Galileo. Certainly, Galileo never mentioned him in his notes. Benedetti, like Leonardo da Vinci, should have influenced scientific development but, apparently, did not.

Galileo, of course, understood and demonstrated that falling bodies have a constant acceleration unless air friction slows them down. His arguments on the subject are worthy of study. For example, he discussed a ball cut in half. There was no disagreement that the two halves, being of the same weight and material, would have to fall at the same rate and time. Well, Galileo asked, what would happen if the two halves were connected by a very light, strong string? Would they fall at the rate determined by the weight of each half or by a rate determined by the weight of the whole? If we answered that the rate would be that of a half ball, then Galileo would ask, "What difference would the length of the string make?" Then he would propose to shrink the string to nothing. Would the rate still be that of a half ball? The only answer that makes sense is that a ball and a half ball fall at the same rate. Hence, we conclude with Galileo that Aristotle was wrong. Object fall at a rate determined by gravity, the same for all objects so long as air friction is negligible.

We have already noticed many of Galileo's contributions to ideas of and knowledge about motions. He was first to grasp the basic nature of projectile motion as a combination of vertical free falling and simultaneous horizontal constant speed movement and he was the first to show by actual measurements that free fall is uniformly accelerated motion. With that understanding in hand, he was the first to see that projectiles follow parabolic trajectories.

One more important advance made by Galileo probably grew out of his free fall experiments with balls rolling down inclines. He realized that if the angle of incline was set to zero, the balls would continue rolling unless stopped by friction. He concluded that the balls had inertia, an inherent tendency to resist change of motion. Unlike Buridan's impetus which was "permanent" but could be given and taken away, Galileo's inertia was inherent and associated with change in motion rather than with motion itself. It seems his idea of inertia was that it continued circular motion since horizontal motion on the surface of the Earth is really motion along a circle. Even Galileo could not break away from the idea of circular motion as somehow natural and perfect. All these advances where made without reference to the causes of motion. Galileo, like the Mertonians, had learned the advantages of Ockham's hair-splitting distinctions.

To be strictly accurate, Galileo never used the word "inertia". That word is Latin for "laziness" and, as such, was the word Kepler used to explain planetary motion. Kepler realized that if a force of attraction from the Sun holds the planets in orbit, then there must also be a tendency within the planet to resist the attraction (since the planet is not actually pulled to the Sun). He called this tendency "laziness" because he thought of it as a simple tendency toward a state of rest.

Figure 20. A demonstration that forward motion continues as Galileo said. When the train is stationary on the left, the ball is in free fall. When the train is in motion, the ball retains the forward motion while in free fall and becomes a projectile on a parabolic path.

Galileo made a further contribution to the understanding of motion with the discussion of a thought experiment in his great book *A Dialogue Concerning the Two Chief World Systems*.[xv] He considered a sailor dropping a stone from the top of a mast of a ship as it slid past an observer on a wharf. In direct contradiction of the Aristotelians (and particularly of Franciscus Capuano of Manfredonia who wrote in 1493 that the stone would fall behind the mast), Galileo insisted the sailor would see the stone fall straight down to land at the foot of the mast. To the observer on the wharf, however, the stone would fall forward as well as down since its inertia would prevent it from losing the forward motion it already had through its connection with the moving ship. Ever the mathematician (and first modern scientist), Galileo quickly proved the path must be parabolic. Thus, we learn the shape of the path of a projectile but we also see that projectile motion is free fall motion plus horizontally constant motion. Finally, we can see that the shape of the path of the stone is relative to the motion of the observer. An observer (the sailor) moving with the stone horizontally will see only vertical, free fall, motion. An observer with a different horizontal motion will see the stone fall on a parabolic path. We call this difference "Galilean Relativity."

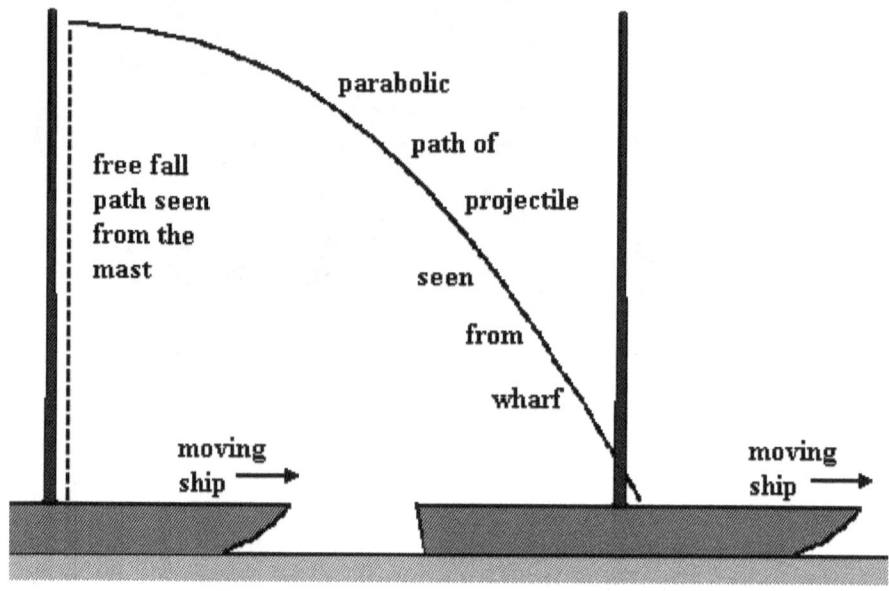

Figure 21. Galileo's thought experiment. From the wharf, the path of the stone is parabolic although the sailor who dropped it from the mast sees simple free fall to the foot of the mast.

Descartes found Galileo's lack of interest in causation unsatisfactory. Like Aristotle, he abhorred the idea of a vacuum, seeing the universe as a plenum, chock-full of matter with no empty spaces left over. Causal explanations for Descartes had to be mechanical accounts of collisions of bodies with other bodies or with media. He was unable to develop these ideas, however, and a fully causal theory of motion eluded him. Possibly because of his strong bias in favor of geometric shapes, Descartes was the first to understand that inertia involves only straight-line motion. The curvilinear impetus of Buridan and the circular inertia of Galileo, Descartes rejected.

His concept of the universe as a plenum originally set in motion by God gave him a glimpse of what would become a major insight. The inertia of bodies kept them in motion and, since the motion had continued for a long time, obviously the inertia or quantity of motion had not diminished. In other words, there must be some quantity of motion that is conserved or constant. Because he saw matter as fundamentally passive and forces as occult concepts to be rejected, he was not particularly clear on this quantity of motion but he apparently had in mind the product of the size of the body and its speed, something very like Buridan's impetus. Thus, Descartes approached a concept of the conservation of momentum.

The Dutchman Christiaan Huygens introduced further subtlety into ideas of motion with his investigation of the concept of *center of mass*. For example, in the collision of two objects viewed from a stationary position, one or both are in motion before the collision. After the collision, the motion of each is changed both in speed and direction. Huygens suggested viewing the collision from a particular moving point (the common center of mass of the objects), the collision would appear, from that point, as merely reversing the directions of the motions, leaving the speeds unaltered. In modern language, a collision viewed from the center of mass (reference) frame is greatly simplified compared to its appearance in the laboratory (reference) frame.

In mathematically showing that this result must hold in collisions, a necessary assumption is that the quantity mv^2 is conserved (m is the mass and v the speed of a body). That is, the sum of this quantity for the two objects before the collision must equal the sum after the collision. Like Kepler's second law, Huygens' discovery pointed in a new direction for physical science, the direction of focusing on conserved quantities. Leibniz named this quantity *vis viva* or "living force" (this was, by way of contrast with *vis mortua* or "dead force", the name he gave the weight of a body). Huygens also applied Descartes' concept of inertia as a linear tendency to the problem of centripetal force, and was thereby able to derive the result that centripetal force has the form mv^2/r where r is the radius of the circle of motion.

All these slow developments came together and were synthesized into a coherent whole by Isaac Newton. Summarizing years of his own work and thought, as well as that of others, Newton proposed a new physics based on new laws of motion. In his major work, The Mathematical Principles of Natural Philosophy (the **Principia**), Newton began with these definitions:

1. "The quantity of matter is the measure of the same, arising from its density and bulk conjointly." (mass = density x volume)

2. "The quantity of motion is the measure of the same, arising from the velocity and quantity of matter conjointly." (momentum = mass x velocity)

3. Inertia is "a power of resisting, by which every body, as much as in it lies, continues in its present state, whether it be of rest, or of moving uniformly forwards in a right (straight) line."

The second definition is almost the same as Descartes'. Newton also echoed Descartes in regarding inertia as pertaining only to linear motion. The insight that inertia keeps a body at rest if it is initially at rest seems to have been Newton's alone. With these definitions as background, he proceeded to enunciate his three laws of motion.

1. "Every body continues in its state of rest, or of uniform motion in a right line, unless it is compelled to change that state by forces impressed upon it."

Paraphrased, the law might be that momentum is constant in amount and direction unless a net force acts. Quantities having direction as well as amount or magnitude are called vectors. Velocity and momentum are vector quantities.

2. "The change in motion is proportional to the motive force impressed, and is made in the direction of the right line in which the force is impressed."

In more modern terms we say the net force equals the rate of change of momentum. If the mass of the object does not change this law takes the familiar form $\mathbf{F} = \mathbf{ma}$ where the boldface type implies the quantity is a vector.

3. "To every action there is always opposed an equal reaction: or, the mutual actions of two bodies upon each other are always equal, and directed to contrary parts."

There are many subtleties in these laws and definitions. In the first two laws mass plays the role of a resistance to change. It is, in fact, the measure of the inertia of a body so we call it the "inertial mass". In the law of Universal Gravitation, mass plays a different role-it is the source of gravity. Newton regarded these obviously distinguishable quantities as the same! Why this should be true at all is a deep mystery that served as the springboard for Einstein in constructing his General Theory of Relativity. Einstein concluded that they are the same because inertial mass creates gravitational attraction by warping space itself. The warping is proportional to the amount of inertial mass producing it.

The first law directly contradicted the Aristotelian law that an object ceases to move when its impelling force ceases. For Newton, thinking in terms of motion in a resistanceless medium, motion was continuous once forces cease to act. For Aristotle, all motions on Earth were motions in a resisting medium in which motion must stop if the force causing it ceases to act. Such simple changes in terms and models can lead to very great differences in physical perspective and the laws required to suit that perspective. That is not to say that either view is right. New information on projectiles, for example, was incompatible with Aristotelian views but was consistent with Newtonian physics. Plausible views that fit most of the facts cannot be true if they cannot fit new, more carefully obtained information.

Newton's statement of the second law is actually defective. About 1750, Euler corrected it to begin "the *rate of* change in motion is...". Newton handled the law ambiguously, treating force sometimes as the change of quantity of motion and sometimes as the rate of change of quantity of motion. This imprecision led to a long-standing state of confusion over the best definition of force into which questions of the conservation of quantity of motion and *vis viva* intruded.

The third law seems to spring full-grown from the mind of Newton, as did Athena from the mind of Zeus, unless Kepler's recognition that a falling body attracts the Earth was a precursor. It is surely the subtlest of all the laws. Note that two bodies are necessary for the existence of forces and that forces always come in pairs. One force, the action, acts from the first body onto the second. The second force, the reaction, acts from the second body onto the first. It is equal in magnitude to the first and acts in the opposite direction. An action-reaction pair of forces never act on the same body. One acts on the first body, the other on the second body. While it frequently happens that two equal and opposite forces act on a body, such a pair can never be an action-reaction pair.

The second and third laws combined require that the total momentum, the total "quantity of motion", be constant for any event. Thus Descartes' vaguely grasped insight eventually found expression as one of the major conservation laws of physics-the total momentum of the universe is unchanging. We have a new answer to the old question: "What is constant in a world of change?" That is, "The total momentum of the universe never changes."

Newton also made important remarks about the nature of space and time.

> Absolute, true and mathematical time, of itself, and from its own nature, flows equably without relation to anything external..." As to space, "absolute space, in its own nature, without relation to anything external, remains always similar and immovable. Relative space is some movable dimension or measure of the absolute spaces...

Leibniz, Newton's contemporary and competitor, made the very modern sounding objections that absolute space cannot be observed and an absolute time, unrelated to anything in the universe, makes no sense. Combined with the calculus and the new gravitation laws, these three laws of motion became the new physics that rapidly replaced the discredited physics of Aristotle.

The next 200 years after Newton saw the gradual unfolding of the implications of Newtonian concepts of motion. Success at explaining more and more things is the ultimate test of scientific theories. Newtonian physics created a record of success superior to anything previously imagined. As more experiments were done and as more predictions were confirmed, Newtonian physics began to seem beyond contradiction. Since no exceptions to Newtonian physics were known, Einstein's 1905 claim that the Newtonian laws were defective came as a great shock.

Only one important critic of Newtonian physics was active between Leibniz and Einstein, Ernst Mach (1838-1916). Like Leibniz, Mach refused to believe in absolute time or space. Both of them saw time and space as relational entities that must be relative to whatever it is that allows us to detect and measure them. Mach also pointed out the circularity of the Newtonian definition of mass; mass is the product of density and volume and density is the ratio of mass to volume. To eliminate this defect and to simplify the laws of motion, Mach recast the third law of motion into a definition of mass; that is, for two interacting bodies, the ratio of the accelerations gives the ratios of the masses. The masses are *assumed* constant. This assumption has to be scored as a major failing of a scheme specifically designed to *define* mass. Rather than eliminate the defect Mach relocated it.

Mach's ideas of relational space and time led him to the interesting and influential position of considering the possibility that inertia is not an inherent property of a body but is due to the distribution of mass outside the body. Einstein later elevated this thought into a principle, the Mach Principle, and he tried to make it part of the foundation of General Relativity. In this he did not entirely succeed.

The Twentieth Century-Relativity

One way of looking at Einstein's work in Special Relativity (which, ironically, is more a form of absolutism than a form of relativism) is to think in terms of the previously mentioned ancient Greek question, "What is constant in this world of change?" Answering, Einstein decided the speed of light and the mathematical forms of the laws of physics must be constant for all observers, no exceptions allowed. So, there are indeed some things that are not relative to the motion of the observer.

Einstein's decision was not as arbitrary as it may seem. By floating a very sensitive optical instrument in a pool of mercury (to eliminate mechanical noise and allow for extremely smooth rotation of the apparatus), the Americans Albert Michelson (1852-1931) and Edward Morley (1838-1923) attempted in 1887 to detect the motion of the Earth through space. They found no change in the light coming to them regardless of the direction from which it came. The Michelson-Morley experiment very nearly forced the conclusion that the speed of light is constant, regardless of the motion of the observer.

The constancy of the laws of physics is a metaphysical assumption, but a very plausible one. Taking these assumptions as a starting point, Einstein found that more things are relative than was previously thought. We have seen that the shape of the path of a projectile is relative to the motion of the observer. Einstein added mass, length in the direction of motion and time to the list of relative quantities. Newton's absolute time prove elusive, just as Leibniz had complained.

Einstein was able to give precise formulas for how these quantities should be altered by motion. To a stationary observer, time on a moving body appears to slow down and the length of the body parallel to the motion appears to shrink. Conversely, the mass of the moving body seems to increase. Each change depends on the speed of the body so that the appearance is relative to the relative motion between the body and the observer, just as Galileo had previously recognized that the apparent path taken by a moving body depends on the relative motion between the body and the observer.

These ideas entail some very strange predictions. For example, suppose a train 100-m long is passing through a tunnel 50-m long at the impossible rate of 94% of the speed of light. According to Einstein's equations, observers in the tunnel will see the train as 33-m long and the tunnel as 50-m long. Consequently, they will agree that there was a time when the entire 33-m of the train was inside the tunnel! Also according to the equations, observers on the train, for whom the tunnel is moving, see a 100-m long train and a tunnel less than 17-m long. These observers will all claim there never was a time when the entire train was in the tunnel! Who is right? Both are stating the case as they saw it. Thus, the seeing of the entire train contained in the tunnel is another of the things that is relative to the motions of the body and its observer.

Newton's fundamental error was assuming time is the same for all observers (absolute time). Einstein made the problem clear with a wonderful thought experiment (Figure 22). A train is traveling very fast with observer B riding exactly at its middle. Just as B reaches observer A, who is stationary beside the tracks, bolts of lightning strike the front and back of the train from observer.

A's point of view. That is, at that instant, the front and back of the train are the same distance from A so light rays from the two events, traveling at constant speed, reach A simultaneously. However, B is traveling toward the front so light traveling from the front event will reach him sooner than light from the back. Light rays from the back event will have to travel farther to reach B. Hence, B will not see the events as simultaneous but will see the front event first. Einstein pointed out that simultaneity is relative to the motion of the observer and the events. Thus, time measurements (and time itself) are relative to the motion of the observer and the observed events.

How was it that Newton and others had gotten things wrong? Einstein's results for bodies at low speeds (less than about a tenth the speed of light) are indistinguishable from those predicted by Newtonian theory. "Relativistic" effects are noticeable only at high speeds, at speeds, in fact, greater than those anyone is familiar with personally. Hence, our intuition, built on our personal experience, is fundamentally Newtonian. That is why these relativistic effects strike us as so peculiar (even impossible). We continue to teach and use Newtonian physics for most situations because it is so much simpler than relativistic physics. Also, the calculational errors of Newtonian physics cannot be noticed except for objects traveling at speeds greater than about 90% of the speed of light.

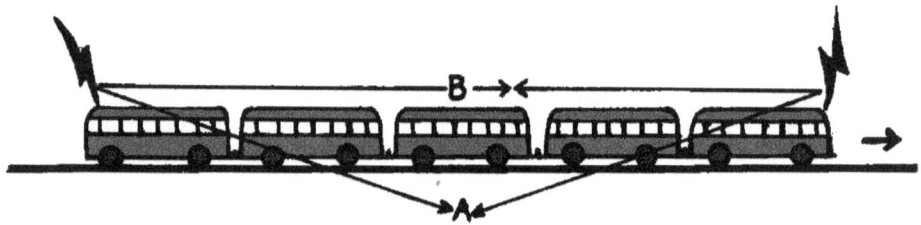

Figure 22. Einstein's Thought Experiment. Lightning strikes both ends of the train simultaneously, as A sees it. B does not agree.

Certain strong convictions and attitudes underlie Einstein's work. His famous remark "God does not play dice" was his cryptic way of stating his deeply held confidence in a completely deterministic, highly ordered universe. He is also on record affirming that "God" to him was Spinoza's God (to over-simplify, "Nature"). Einstein attributed his courage in challenge the authority and tradition of Newtonian physics to his early rejection of authority. In his early teens, apparently, he became convinced that religious, social and political authorities were suspect. He claimed this suspicious attitude was important in enabling him to consider and make effective use of new ideas.

Special Relativity was *special* in that it only considered things from the view of unaccelerated observers moving at constant speeds with respect to each other. Most importantly, Special Relativity does not work for observers in a gravitational field because they are accelerated. Einstein disliked this limitation and quickly set about removing it. After more than a decade of work he succeeded, finding a view of relative motions we call the General Theory of Relativity, or General Relativity.

The starting point of the theory is the observation that, over small regions of space, gravitational attraction cannot be distinguished from an accelerated state. Einstein used the example of an observer in an elevator (no windows allowed!). If the elevator is falling in a gravitational field, an object dropped by the observer remains apparently stationary because it is falling along with the observer. On the other hand, if the elevator is accelerated upward in a region where gravity is negligible, the observer will be pressed into the floor of the elevator just as if gravity were acting to produce sensations of weight. These considerations led Einstein to posit that gravitation is an inertial effect. The Mach Principle was interpreted by Einstein to mean that inertial effects must be related to the distribution of mass in the universe. Hence, gravity must relate to the mass in the universe. This connection led him to the conclusion that the presence of mass deforms the space-time web of the universe, creating gravitational attraction for other masses.

With the development of General Relativity, the view of inertial motion has become considerably complicated. Newton's idea that inertia causes objects to travel in straight lines if no net force acts on them must be revised. In General Relativity, an object on which no net force is acting travels on a geodesic line (a line of shortest distance) in space-time. In a region where there is no other matter, the geodesic line is a straight line. Near a massive object, geodesic lines are curves. In fact, a planet in orbit about the Sun is also travelling on a geodesic line of space-time. If it is no longer clearly true that inertia is inherent in an object, neither is it completely false. In fact, there seems to be no current consensus among physicists as to just how inertia and inertial motions are to be understood.

Further Reading

Galileo's Discovery of the Parabolic Trajectory, Stillman Drake and James MacLachlan, Scientific American, March, 1975

Einstein, The Life and Times, R.W. Clark, The World Publishing Co., 1971.

In the Presence of the Creator Isaac Newton and His Times, Gale E. Christianson, The Free Press, 1984.

Reference Systems and Inertia, Beryl E. Clotfelter, The Iowa State Univ. Press, 1970.

"The Science of Motion", John E. Murdoch and Edith D. Sylla in *Science in the Middle Ages*, ed. David C. Lindberg, Univ. of Chicago Press, 1978.

Chapter V

Matter and Atoms

The Classical Period

In the classical Greek period there were two main views of matter. Matter was either regarded as a continuum or it was seen as atomic, made of discrete, irreducible parts. In both views matter came in varieties. The varieties were explained by assuming all materials are combinations or mixtures of basic types of matter called elements. Elements are not to be confused with atoms. For the Greeks, elements were the materials from which all other materials were made, the unchanging features of a changing material world. Atoms were the irreducible, invisible particles of which any sample of an element is composed. Although we now know the atomic structures, we retain the same basic distinction between an element and its atoms. Typically, those with a continuum view of matter believed there were four (or fewer) elements. The atomists, on the other hand, usually assumed a very large or infinite number of elements. In both views, matter occupied space.

Empedocles of Sicily (d. 434 BC) was the first to abandon the Ionian hope of finding a single basic element and to claim instead the four elements earth, air, fire and water. Everything, he said, is made of mixtures of the four elements. The two forces of love and strife do the mixing. The former holds things together and the latter effects the changes in objects through its tendency to cause dissolution. His cosmos was made of two hemispheres, day and night, and the Moon shone by reflection of sunlight. He believed light has a finite speed.

Perhaps in response to the paradoxes of Zeno, the materialist and early atomist Anaxagorus denied that matter can be divided to the point where each piece is nothing. The "seeds" of matter limit divisibility. For him everything was made up of mixtures of imperceptibly small seeds of various types. The totality of matter was conserved because the seeds cannot be destroyed. Changes in matter reflected changes in the seed mix.

The great atomists were Leucippus of Miletus and his student Democritus of Abdera (b. 495 BC) followed by Epicurus (341-270 BC). While each had a somewhat different view, there was substantial agreement among them. Leucippus and Democritus were the first to use the word "atom" (atomos = uncuttable). For all of them, the universe consisted of invisible atoms in the void, jostling, colliding and cohering atoms making compounds and being knocked out of compounds in a seething, ceaseless turmoil. Motion and change were the normal state of the universe. It was not change but constancy that required explanation. Even sensations and the mind were explained atomically. Materialistic explanations of the mind are not new.

Atoms were irreducible and, therefore, permanent so the number of atoms was conserved. As Lucretius said, presumably following Epicurus, "nothing comes from nothing". At this point, the Greek atomists came very close to the idea of conservation of mass. The atomists disagreed on the numbers of types of atoms, Epicurus thinking the number finite, the others holding out for an infinite number of types. The cosmos was infinite, as was the total number of atoms. Evidently, all the atomists were functional atheists, relegating the gods to outside the universe and arguing the imperfections of the world make any divine origin implausible. Their view of the universe was mechanistic, as opposed to the animistic views of other philosophers of the ancient world.

The atomists dealt with the problem that matter appears to be continuous and solid by assuming that atoms in contact could hold to each other by hooks and velcro-like protrusions. Their shapes distinguished the types of atoms. Atoms of liquids were smoother than those of solids since solids needed more protrusions on the atoms to make the strong bonds needed to give solids their hardness. Atoms of soul or mind were the smoothest of all, accounting for the elusive nature of their substance.

Plato made a radical break with tradition in that he combined and synthesized the two views, retaining both atoms and the four elements. In the *Timaeus*, he presented a geometric, atomic theory of the four elements. The atoms of fire were tetrahedral, those of earth were cubic, of water, icosahedral and of air, octahedral. These "Platonic solids" were four of the five regular solids recognized by Theatetus, the last being the dodecahedron (Figure 2). All but the dodecahedron and the cube have sides made of equilateral triangles. The sides of the cube are, of course, squares and those of the dodecahedron are pentagons. Equilateral triangles and squares can be cut into two right triangles but pentagons cannot be so divided. Plato saw these differences as a way of distinguishing and separating the dodecahedron from the other four shapes. The fact that there were four such shapes and only four elements on Earth was too strong a conjunction for Plato. He felt constrained to see the conjunction as proof of a deep connection, a connection he placed at the atomic level.

A consequence of the criteria for eliminating the dodecahedron was that his atoms were not made of equilateral triangles and squares but of half-equilateral triangles and half-squares. These half-figures played the role in Plato's theory that electrons, protons and neutrons play in today's theories. The Platonic atoms were not indivisible and permanent like the atoms of Democritus. They could be disassembled into their constituent half-figures and reassembled into atoms of different elements. Since the half-squares are not the same shape as the half-triangles, earth (the cube) could not be made into the other elements. Plato allowed for variety within the theory by assuming the figures came in different sizes so, for example, there were different earths made of cubes of different sizes. Of course, the atoms were too small to be visible. Plato seems to have thought matter was eternal.

The teleological system of Aristotle in which all natural change aimed at reaching a goal made possible a rich and subtle view of matter. The universe was a "plenum", packed full of substance without any voids or empty spaces. Nature abhorred a vacuum. Thus, Aristotle sided unambiguously with the proponents of matter as a continuum. Unqualified matter without form did not exist in the Aristotelian universe although, as a logical category it could be discussed as something with the potential to take on form. When the potential was realized, the unqualified matter became substance ("actualized" matter).

It was the goal of matter to perfect itself by assuming a form. The form assumed was regarded as the formal cause of the change, one of the four causes of any change from potential to actual form. For Aristotle as for Plato, the elements could transmute. Air, for example, had the potential to be transformed into fire. Like Plato, Aristotle seems to have thought of matter as eternal. Neither of them had a concept of conservation of matter nor were they able to see mass or weight as a characteristic of matter because aether did not respond to gravity but was material nonetheless.

The Aristotelian focus on goals and causes of goal-directed change and on the potentialities and actualizations of material, made atomism irrelevant. The microscopic details supplied by atomism contributed nothing to the understanding of goal-directed behavior. The differences ran even deeper, however. Mechanical laws and chance events ruled the universe of the atomists. The Aristotelian universe was ruled by order and its pattern was of orderly change from potential to actual forms. Aristotle strongly opposed atomism, not because atoms themselves were problematic but because atomistic theories were metaphysically unacceptable.

The Stoics, Zeno of Cition (332-262 BC), Chrysippus of Soli (280-207 BC) and Poseidonius of Apamea (135-51 BC) made the inert, passive plenum of Aristotle an active continuum constantly held together by *pneuma*. The word literally means breath or spirit but it also means air. In fact, air was the sense in which the Stoics originally used it but the concept evolved into something that was a combination of air and fire and, as such, it came to resemble Aristotle's aether. Both were thought to be the medium filling all space which otherwise seemed empty. The Stoics thought *pneuma* also pervaded matter. The amount of *pneuma* in a material and the mix of fire and air in the *pneuma* determined the characteristics of the material.

Eventually, the *pneuma* became a means of sustaining harmony and relatedness throughout the universe; that is, it evolved into a universal *pneuma*. This view underlies Poseidonius' idea that a sympathy with the Moon and Sun caused the tides. Of course, *pneuma* was associated with *psyche* (soul) and its binding of matter together was attributed to "forces" so the word force came to have associations with spirit and soul, the "occult" connections Descartes so much despised. This universal *pneuma*, the best *pneuma*, the *pneuma* composed essentially of pure fire, became almost deified into a supreme god.

The Interlude

Shorn of its materialism, the cosmic *pneuma* easily transformed into something resembling the Christian God. Hence, taken together with the ideal forms and later the teleology of Aristotle, the continuum view of matter was the preferred view for many centuries. The determinism and anti-supernaturalism of atomism prevented its serious consideration in a Christian framework.

There were two main developments in the understanding of the nature of matter during the Interlude. Alchemy arose in Alexandria and came to dominate much of the thinking about matter for a millennium. It grew from the ideas about transformation of materials in classical Greek thought which, in the hands of the alchemists was altered into a hope of making gold from cheaper metals. Other classical ideas also were changed. For example, Theophrastus Philippus Aureolus Bombastus von Hohenheim of Basle (1493-1541), who perhaps not surprisingly preferred to be known as Paracelsus, considered matter to be composed of three principles: sulfur, mercury, and salt bound together by an organizing principle he called *Archeus*. His three principles were not actually the materials sulfur, mercury, and salt because materials always contained *Archeus*. Interestingly, the idea of mercury and sulfur as principles in materials derives from the rediscovered works of Aristotle.

By and large, however, alchemy made few theoretical advances because it was not theoretical in focus. Whether by fair means or fraudulent, the alchemists all intended to get practical results. And practical results were obtained, although the ultimate goal of transforming base metals into gold was never reached.

Alchemy advanced the techniques of chemical processing and the knowledge of new materials. Around 1270, Raymond Lully discovered nitric acid and Basil Valentine reputedly discovered hydrochloric acid. The Arab alchemists were the first to see alcohol in a highly purified state. The discovery of liquids plainly distinct from water eventually weakened the idea of water as an element. There were also discoveries of new solids; Paracelsus was the first to recognize zinc as metallic, Valentine discovered antimony and, in 1669, Heinrich Brand discovered phosphorus while experimenting with urine.

A large body of metallurgical knowledge developed, although it was largely disconnected and unorganized information. In fact, the alchemical writings of the Shi'ite Moslem, Abu Musa Jabir ibn Hayyan were so characteristically incomprehensible that his name has become an English synonym for the incomprehensible. Jabir was known in the West as Geber and his writings were *gibberish*.

Most of the attention and technology in alchemy focused on solids and liquids, especially metals and acids. Little work was done on gases. Water vapor was not distinguished from air nor air from products of combustion or respiration. In short, gases were confused. Advances here had to wait until Jan Baptista van Helmont, in the 17th century, noted that respiration involves an exchange of "gas" in the lungs. He distinguished several types of gas: *gas carbonum* (CO), *gas pinque* (CH_4) and two types of *gas sylvester* (CO_2 and NO_2).

The second important development was transubstantiation, an orthodox doctrine of the Eucharist. Although it arose earlier, Aquinas gave it its basic form in the remark "all the substance of the bread is transmuted into the body of Christ..." By substance Aquinas meant the Aristotelian concept of informed matter, matter in a specific form (species). Aristotle's view of matter was that all bodies had a fundamental substance, uninformed matter, plus form, those secondary qualities (of activity and appearance) called species. Thus, matter had qualities but could also be quantified because it takes up space (has extension) and the size can be measured.

Now, the Host never changes its appearance; it always appears to be bread, even after consecration. That, Aquinas said, is the essential miracle and mystery. The matter of the substance becomes separated from its extension, the species separated from its proper substance. The appearance of bread separates from the substance of bread and attaches to the substance of flesh. Appearance is accidental to substance so the remark arose that in the consecrated Host one has "accidents without subject". When Thomism became the official theology, this became the orthodox view of communion. The important point is that a major religious doctrine became tightly bound to the Aristotelian view of matter.

Nominalists, like Ockham, disagreed. For them, substance was inseparable from extension (the occupying of space). Since extension is measurable, substance is quantifiable (*res quanta*, a quantifiable thing). Furthermore, taste, color and smell of things, the accidents of the Aristotelians, were also quanta and inseparable from substance (*substantia est quanta*). The miracle of the Eucharist, in the nominalist view, was that substance is separated from its real characteristics. This may sound like hair-splitting, but the leaders of the church saw it as a threat to orthodoxy and orthodox authority and they condemned it (and Ockham) for a while.

Following his fellow Franciscan, Roger Bacon (and his mentor, Grosseteste), Ockham inclined toward atomism, espousing the view that heat is transported by a dispersion of particles. Nicole d'Autrecourt also combined nominalism and atomism. The appearances of materials he attributed to positions and motions of atoms. He even suggested "light is nothing else but bodies" and the apparently infinite speed of light is, in reality, simply a very great but finite speed. John Wycliffe in England and then John Hus in Bohemia also espoused nominalist views of matter, in addition to their activities as religious reformers. Luther, educated at the nominalist dominated University of Erfurt, rejected transubstantiation and advocated consubstantiation where the substance of both bread and the body of Christ are present in the Host at the same time.

The Scientific Revolution

The Counter-Reformation began with the Council of Trent. In 1551, the Council reaffirmed the orthodox view of transubstantiation. From that time on, advancing atomism or other non-Aristotelian views of matter became an activity likely to draw the attention of the Inquisition. Galileo, however, dared to advance atomistic theories, first in *The Assayer* and later, less openly, in his *Dialogue*. In *The Assayer*, he described heat as a substance (recall the element fire) produced by the breaking up of a body into infinitesimal parts. This process was carried out through the action of particles he called "fiery minims" which he distinguished from the infinitesimal parts of the body. This use of words conformed to late Middle

Ages usage. The Aristotelians thought a material could be divided only so far before the next division produced pieces that no longer had the attributes of the original material. These minimum pieces were called *minima*. Galileo was apparently trying to seem Aristotelian still.

The word "atom" Galileo used only for particles of light. He plainly embraced nominalist attitudes toward universal forms and secondary qualities but he carefully avoided any mention of Eucharistic theology. Interestingly, Pietro Redondi has recently argued that it was for his atomistic views, not for his Copernicanism, that Galileo was subjected to the Inquisition.[xvi]

Looking for safe ways of supporting atomism, Galileo's students turned to the study of air. If atoms were not a safe topic, perhaps the vacuum, the reverse side of the atomism coin, would be acceptable. Hero of Alexandria (284-221 BC), in his *Pneumatica*, had long before seen the power of the image of "atoms in the void" for explaining compressibility of air. Air is compressed, in this view, by reducing the amount of void separating atoms. Suction pumps were already in use and it was known that such a pump could not pump water up more than about 30 feet. Galileo's students realized that the pressure of the air was the cause of the rise of water and not the suction of the pump. Evangelista Torricelli, Galileo's secretary, suggested replacing water with mercury. Because of its greater density, he expected mercury to rise much less than water. The year after Galileo's death, Viviani and Torricelli took a long glass tube sealed at one end, filled it with mercury and inverted it into a pool of mercury. The mercury in the tube ran out until about 30 inches remained above the level of mercury in the pool. Above the mercury, between it and the sealed end, Viviani saw a vacuum. The Cartesians did not see it that way. They saw the "most subtle kind of matter" above the mercury. It had to be subtle or it could not quickly penetrate and flow through the glass of Torricelli's tube!

Torricelli and Viviani soon noticed that the mercury level fluctuated by small amounts from day to day and rightly attributed this to variations in atmospheric pressure. They had invented the barometer. Blaise Pascal, in

France, quickly predicted the barometer reading should decrease with altitude because a reduced fraction of the atmosphere lies above it. Never robust himself, he persuaded his muscular brother-in-law to carry a barometer to the top of Puy de Dome in central France. As expected, the mercury level declined as the altitude increased.

The vacuum pump was invented by Otto von Guericke (1602-1686), the mayor (burgomeister) of Magdeburg, the town for which his famous hemispheres are named. The hemispheres, assembled as a sphere, were easily separated unless the air inside was removed by an air pump. Teams of horses could not then separate them. By evacuating glass vessels, von Guericke found that light can traverse a vacuum but sound cannot. He also found that animals die in a vacuum.

Figure 23. Viviani and Torricelli and their barometers.

In England, Francis Bacon defended atomism at first although he later repudiated it. In the essay "Thoughts on the Nature of Things", he used the example of a bit of saffron. Warning readers not to be so foolish as to mistake for atoms the small pieces one can see when examining the powdered saffron, he asked how it is that a small amount of saffron can turn water uniformly yellow? The answer, he suggested, can only be that the "atoms" of saffron can fill the empty spaces between the "atoms" of the water. It remains, even today, a very suggestive demonstration.

One of the earliest and most important followers of Francis Bacon, Robert Boyle had an able assistant in Robert Hooke. Using a pump designed by Hooke, Boyle launched a lengthy and systematic research program on the vacuum and air pressure. He found that flame goes out, insects cannot fly and smoke falls rather than rises in a vacuum. He confirmed von Guericke's observations that animals die in a vacuum. He also noticed that a piece of "roasted Rabbet" was preserved for several months in a vacuum. Boyle's "rabbet" thus has the distinction of being the first "vacuum packed" food.

Turning to the study of air at varying pressures, he found that the volume (V) of the air varies inversely with the applied pressure (P, at a fixed temperature). The relationship, $P = C/V$, is known as Boyle's Law where C is a constant. As a good Baconian, Boyle arrived at this law by making many measurements and extracting the law inductively from the data. Independently, Edme Mariotte (1620-1284) obtained the same results by an hypothetical-deductive method more in keeping with his Cartesian presuppositions. Mariotte began by assuming the inverse relationship, then confirmed it by a few careful measurements. On the continent of Europe, Boyle's Law is frequently called "Mariotte's Law".

Boyle and Mariotte differed in interpretation as well as method. Boyle saw the behavior of the air as a confirmation of atomism with atoms of air acting rather like Hooke's springs. Mariotte gave a typically Cartesian, continuum explanation. To this point, there was evidence for the existence of a vacuum and atoms but none of it was conclusive or above dispute. In

1666, Boyle set out to improve the status of atomism with the publication of *The Origins of Forms and Qualities*. Like Bacon, Boyle saw matter in terms of corpuscles in motion in a vacuum but his idea of atoms was far from modern. He spoke of corpuscles "simple, compounded or decompounded...so firmly united that they will not be totally disjoined, or dissipated,...by...fire or heat", and believed that matter is conserved. Yet, in his *Sceptical Chymist* of 1661, he allowed for atoms to have virtually unlimited transformations and combinations and remarked of elements, "I see not why we must needs believe that there are any primogeneal and simple bodies, of which, as of pre-existent elements, nature is obliged to compound all others".

Boyle's differences with modern ideas stem from difficulties that were almost insurmountable in his time. Firstly, atomism still had very strong atheistic overtones and Boyle, the devout Christian, was very cautious about clearly advocating it. Also, chemical techniques for identifying and separating elements were almost non-existent because no one was even sure what characteristics should be used. Taste, color and smell were regarded by atomists as secondary characteristics arising from the impression the material makes on the senses rather than primary characteristics inherent in the material itself. Even the atomists, however, could be misled into identifying materials by colors. The situation was further confused by the ancient identification of fire as an element. What characteristics does fire have in common with matter? We would dismiss the question as a misunderstanding. Thinkers at the end of the 17th century did not have the confidence about this that we would have today, so they were forced to confront the problem of finding common characteristics where there are none (as we now know).

Mass, which seems so obviously important to us as a characteristic of matter, was not well defined until Newton had done his work. It took chemists a long time to see that matter must have mass and that mass conservation was a major principle for understanding chemical behavior. The confusion over mass was compounded by the failure to see air as an active agent. Boyle heated metal in air until, as we would say, it oxidized into calx. Weighing the calx, he found it weighed more than the original metal. Repeating the experiment with a known weight of metal in a sealed vessel, he found the calx, extracted from the vessel, again weighed more than the metal. He concluded that the fire had united with the metal to produce the calx, increasing its weight and changing its composition. Had he weighed the sealed vessel with its enclosed material just before and after heating, he might have come to a quite different conclusion. He never seems to have realized that something in the air itself might have combined with the metal.

In France, Jean Rey had done much the same experiment in 1630 but had weighed the sealed vessel with its contents before and after heating. He had realized that the increased weight of the calx was due to something in the air. Once again, we see different interpretations of the same information. Rey and Boyle agreed, however, on the importance of weight as a characteristic to be traced throughout a chemical process. Thus, weight became a key characteristic of matter and an important step was taken toward clarifying the nature of matter.

Adding an appendix of queries to the 1706 edition of his **Opticks**, Newton placed his enormous prestige and fame squarely behind a mechanical, atomic theory of matter. Although he did not call himself an atomist and never used the word "atom" except in explicit reference to the atomists, his views had admitted affinities to those of the ancient atomists. From the vantagepoint of his own new physics he was able to make critical improvements in classical atomism.

Atoms have mass and, hence, inertia, he argued. They are solid and structureless except that they may be surrounded by a whirling atmosphere of aether. They cohere by a force analogous to gravity or magnetism or even electricity. This force, he imagined, "in immediate contact is exceeding strong, at small distances performs the chymical Operations…and reaches not far from the particles with any sensible Effect". The force was attractive but could be repulsive at times, as, for example, in the expansion of a gas when pressure on it is reduced.

His atoms were "perfectly hard" and completely immutable, their stability guaranteed by God himself. These latter features allowed Newton to assert the conservation of mass because perfectly hard atoms cannot be destroyed. However, both Newton and his contemporaries (and also those who came after them) quite erroneously understood "perfectly hard" to mean "incapable of elastic rebound on collision". Hence, Newton's decision in favor of "perfectly hard" atoms triggered two hundred years of speculation and confusion on the laws of motion for such objects. Since no one could experiment on perfectly hard bodies to resolve the issue, the problem persisted and much time and energy was wasted.

These additions enabled him to explain respiration, Boyle's law, capillary action and various kinds of chemical behavior. Having mass and taking up space had finally become requirements for matter. However, Newton had no clear idea of elements or chemical processes. In fact, his main interest was the nature of light which, like Galileo before him, he mostly regarded as corpuscular although at times he spoke of it in terms of vibrations or "fits of easy Transmission and Reflexions". Evidently he connected these wave-related terms with the medium through which light traveled rather than with the light itself.

For the medium he reconstituted Aristotle's aether. As before, the aether pervaded all space but Newton added new details. Aether must have 4.9×10^{11} times the "elastick force" of air because of the enormous value of the speed of light compared with the speed of sound in air. To get such strength, he reasoned the particles of aether "are exceedingly smaller than those of air, or even those of Light". His idea, evidently, was that the smaller particles could get closer together and experience a greater force, assuming the force, like gravity, increases with decreasing distance from the particles.

Newton's influence had two unfortunate effects: it stopped cold any wave theory of light, and light became a candidate form of matter. In the next century, the only person who dared to challenge Newton's corpuscular ideas of light was someone with almost as great a reputation, Euler. However, Newton greatly extended the range of phenomena made understandable by the atomic theory. With his attractive force and his all-pervading aether, he solved the classical problem of how matter can have the appearance of a continuum and still be made of atoms (although treating aether as atomic rather spoiled the overall effect). Nevertheless, the evidence for atoms was still weak.

The Study of Gases

More than any other substances, gases are suited to an atomic view of matter. The discovery by the late alchemists that there are many, distinctly different gases led eventually to a much-improved understanding of the nature of matter and to the recognition of new elements and new compounds. It was a slow process, however, and beset by confusion. No rules of naming gases existed and different names for the same substance hindered the development of a coherent, rational nomenclature. The distinction between mixtures and compounds was not well grasped because there was no clear idea of atoms, so no clear idea of molecules was yet possible. Nonetheless, terms and ideas were slowly clarified throughout the eighteenth century.

In 1727, Stephen Hales (1677-1761), an Anglican clergyman and Fellow of the Royal Society, published the book *Vegetable Staticks*. Newton gave it the official blessings of the Royal Society as one of his last actions as President. Hales had performed numerous experiments on liquids and gases in living things and in minerals. Throughout, he had been troubled by the difficulty of collecting pure gases unmixed with air. Finally, he learned to bubble gases into an inverted, water filled container. It was simple but a major technical advance nonetheless and it set the stage for later developments. Hales himself collected a gas with his new "pneumatic trough". Calling it "fixt" air, he showed it obeyed Boyle's Law.

Georg Ernst Stahl (1660-1734), working in Germany around 1700, developed the phlogiston theory of chemical processes. Basically an alchemist, Stahl took his cue from Paracelsus and his followers. They had thought of mercury as a "metalizing" principle that made metals shiny and of sulfur as a principle giving solidity and flammability to organics such as wood. Stahl thought the two functions could be due to a single principal, phlogiston. The earlier alchemists explained the calcining of metal saying that heating the metal drove off the mercury and left calx. Wood burned by the application of heat that drove off the sulfur. Stahl simply claimed that phlogiston was the matter driven off in both cases. The phlogiston theory was widely accepted, especially in Germany and France where English (Newtonian) ideas were poorly received but also in England and Scotland. The theory had its peculiar aspects. For example, heating metal supposedly drove phlogiston out of the metal to produce calx but the calx weighed more than the original metal! Phlogiston, its advocates explained, must have levity. Levity! But gravity is attractive! Well then, phlogiston has negative weight!

Joseph Black, in Edinburgh, found in 1753 that Hales' "fixt" air was not the same as ordinary air. He recognized that it was the same as van Helmont's *gas sylvester* which appeared in the burning of charcoal and other processes. About the same time, Henry Cavendish collected the gas given off when zinc was added to an acid. It burned with a blue flame in air and had weight, yet Cavendish identified it with phlogiston. When the new gas was mixed with air in a container through which electric sparks were then passed, a dew formed in the container. Cavendish, working carefully for a decade, was able to show that water is a compound of two volumes of his phlogiston and one volume of what he called "dephlogisticated air". At long last, water was deposed from among the elements.

A dissenting clergyman whose avocation was chemistry, Joseph Priestley, took up the study of dephlogisticated air. Carl Wilhelm Scheele had called it "fire air" because a flame burned so brightly in it. Priestley found that a mouse enclosed in a container of air soon died but one in a container of dephlogisticated air lived much longer. He interpreted this to mean that the dephlogisticated air absorbed phlogiston from the respiration of the mouse and eventually became unfit for breathing. He also found that plants give off dephlogisticated air. Of course, he realized the air must contain some dephlogisticated air normally. Priestley distinguished the gases sulfur dioxide (SO_2), hydrogen chloride (HCl) and ammonia (NH_3) from dephlogisticated air. Air, no longer an element, came to be recognized as a mixture of gases.

The next major figure after Newton in the history of the nature of matter was Antoine Lavoisier (1743-1794). Early in his career as a chemist he resolved to work on a systematic terminology for chemistry. He found himself continually colliding with theoretical difficulties. He wrote:

> While I proposed to myself nothing more than to improve the chemical language, my work transformed itself by degrees, without my being able to prevent it, into a treatise on the Elements of Chemistry.

A Newtonian, Lavoisier began with the conservation of matter. Too cautious to boldly declare that matter must have mass, he decided instead to call anything matter which could be shown to have some conserved quantity about it. Since heat was seemingly conserved, it retained its ancient classification as a material. Lavoisier named this heat matter caloric. He distinguished caloric, the material, from Stahl's principle, phlogiston. A second major decision was the definition of an element. He determined an element must be something chemists were unable to decompose. This criterion gave him a list of eleven elements including light and caloric. Water, earth and air, now recognized as mixtures or compounds, were demoted finally and permanently.

Lavoisier then went to work on the new gases to see if any could be classified as elements. Careful weighing of materials throughout an experiment became a hallmark of his work and he soon knew that calcining metal in an enclosed container did not change the total weight of the container plus its contents. By 1778, he knew that it was Priestley's dephlogisticated air that combined with the metal to create the calx and he recognized that air is a mixture of dephlogisticated air, which he now called "oxygen", and Cavendish's phlogiston or "hydrogen" as Lavoisier renamed it. To Lavoisier, oxygen may not have been exactly a gas in our sense of the word because he soon came to think of it as a principle like Paracelsus' sulfur or Stahl's phlogiston. Combustion and respiration required oxygen. "Earths" (modern "bases") were compounds of metals and oxygen and provided Lavoisier a reason for removing earth from the list of elements. Acids were compounds of non-metals and oxygen in Lavoisier's scheme.

As to phlogiston, Lavoisier complained it was neither clearly defined nor clearly characterized.

> Sometimes (it) is heavy and sometimes it is not; sometimes it is free fire, and sometimes it is fire combined with the earthy element; sometimes it passes through the pores of vessels and sometimes they are impenetrable...

Eventually, it was just this flexibility that made phlogiston unbelievable.

In the final analysis, Lavoisier largely succeeded in establishing a new language of chemistry and laid a foundation on which others could build.

His emphasis on mass and weight also set the stage for a disagreement. The question arose as to whether elements combine by weight or volume. Solids and liquids are best measured by weight but Cavendish and others had shown that gases combine in volume ratios. Claude Louis Berthollet (1748-1822), Lavoisier's coworker, attacked the entire idea of combination by any definite proportion. He was effectively refuted by his compatriot Joseph Louis Proust (1766-1826), a teacher of chemistry in Spain. Proust persuasively advanced the view that elements combine in definite proportions by weight.

The Nineteenth Century

A Quaker and a self-taught meteorologist, John Dalton (1766-1844) was puzzled by the new ideas that air is a mixture of gases of different densities. Yet, as he himself confirmed, the atmosphere has the same composition everywhere. He wondered why the gases do not separate by density. The wondering set him off on a series of investigations. Between the new emphasis on weight and his own concern with the densities of gases, Dalton was led to wondering, as a convinced Newtonian, if the atoms of different gases should have different weights. He decided they should and that the atoms of any particular element should all have the same weight.

Hydrogen is the least dense gas. Accordingly, Dalton arbitrarily regarded its weight as one unit. Taking the same line as Proust, he assumed one atom of hydrogen unites with one atom of oxygen to produce water. Since, by weight, about seven or eight units of oxygen combine with one unit of hydrogen to produce water, Dalton deduced the weight of oxygen as seven units (eight is the better number but, as Sir Humphrey Davy (1778-1829) remarked, Dalton was only a crude experimenter). Looking at the gases produced by combining carbon with hydrogen and water, he encountered more examples of Proust's rule of definite proportions. Ethylene and methane both contain only carbon and hydrogen but methane has twice as much hydrogen for a given quantity of carbon. Also, carbon combines with twice as much oxygen in carbonic acid gas (carbon dioxide) as in carbonic oxide (carbon monoxide).

Dalton realized very quickly that all these results, in fact, Proust's law of definite proportions generally, could be explained easily by assuming the compounds are made by bonding atoms of one element to atoms of another. He built small wooden models of atoms and began to do drawings (Figure 24) of variously labeled small spheres clumped together to represent molecules (which he called "compound atoms"). He developed words to describe compounds. Two atoms bonded together made a "binary compound atom" and many identical such clump together were a "binary" compound, three atoms made a "ternary compound atom" and ternary compound and so on. Although he got many of the weights wrong, the scheme plainly made sense and his ideas were well received. Especially his nomenclature and his models of molecules made Dalton the father of modern chemistry. Incidently, colorblindness is sometimes known as Daltonism because John Dalton had that condition and made one of the early studies of it.

Dalton's atoms differed from Newton's only in slight detail. In particular, the aether atmospheres of Newton were transmuted into caloric atmospheres by Dalton. The basic, hard, solid structure remained.

The first two decades of the century saw light removed from the list of the elements. Through the efforts of the English polymath Thomas Young (1773-1829) and the Frenchmen Augustin Jean Fresnel (1788-1827) and Dominique F.J. Arago (1786-1853), the wave nature of light was convincingly demonstrated in terms of the details of diffractive behavior, interference and the existence of polarized light. It was particularly this later detail which persuaded scientists that light is a transverse wave.

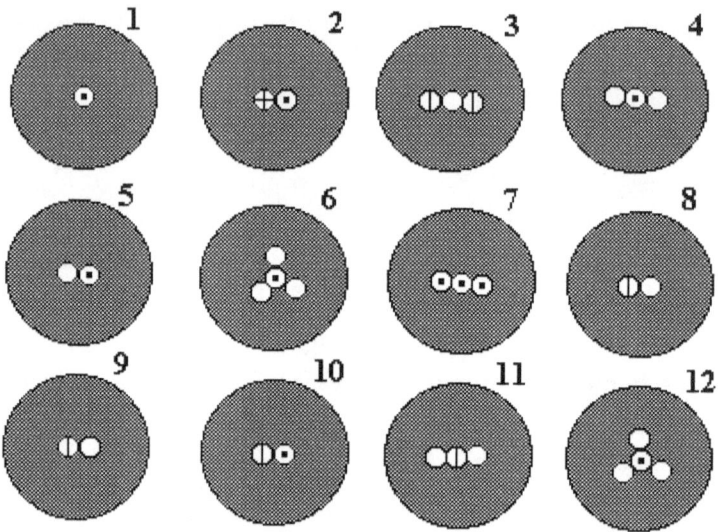

Figure 24. Schematic following Dalton's drawings.of "molecules"-atoms (small circles) surrounded and bonded by caloric fluid (large shaded circles).

Count Amedeo Avogadro (1776-1856) returned to the question of combination by weight or by volume in 1811. Relying on the experimental results of Joseph Louis Gay-Lussac (1778-1850), he suggested that gases combine in fixed ratios by volume because the same number of units of matter (atoms or molecules) occur in a given volume of any gas (at a fixed temperature and pressure). For example, 1 liter of hydrogen and 1 liter of oxygen have the same number of atoms and/or molecules. This number is now known as Avogadro's number and it equals 6.02×10^{23} in 22.4 liters of gas at 0^0C and one atmosphere of pressure.

On the strength of this idea, Avogadro believed 2 atoms of hydrogen combine with one atom of oxygen in water because, as Gay-Lussac (and Cavendish before him) had shown, two volumes of hydrogen combine with one of oxygen to produce water. Thus, oxygen atoms must weigh about 15 times the weight of hydrogen atoms. These numbers were not consistent with Dalton's numbers and led to the idea that elemental gases like hydrogen and oxygen had diatomic molecules. Avogadro's idea was accorded a cool reception. It was forgotten for half a century.

In 1815 was born another idea that took a long time to win acceptance. William Prout (1785-1850), looking at what was known of the relative weights of atoms, became convinced that the correct values really should be integer multiples of the weight of hydrogen. He imagined the heavier atoms all made up of hydrogen atoms strongly bound together. Prout's idea is an example of how a fundamentally sound idea can be derived from incomplete data where more complete evidence would have quenched immediately any hope the idea could be true. Continuing improvements in weighing the elements by J.J. Berzelius (1779-1848) and others made Prout's idea obviously untenable by the 1830's. The striking case of chlorine, for example, which was known by that time to have an atomic weight of about 35.5 plainly was inconsistent with Prout's idea. Of course, the implicit assumption that one atom of chlorine was exactly like every other was not reconsidered. That all chlorine atoms are identical seemed much more likely than Prout's hypothesis. It was the hypothesis that was jettisoned, not the assumption. Nonetheless, a long and largely speculative discussion of the nature and structure of atoms continued throughout the nineteenth century.

James Clerk Maxwell (1831-1879) began to reconsider the theory of gases in the 1850's. Expanding on ideas developed by Daniel Bernoulli (in the 1730's!), Maxwell showed that Boyle's Law and a number of other known facts could be explained by assuming gases to be comprised of many, very small, hard bodies randomly colliding with each other and the walls of their container. It became possible to measure Avogadro's number accurately enough to show Avogadro had been right. There are equal numbers of molecules in equal volumes of any gases at equal temperature and pressure. The success of Maxwell's "kinetic theory of gases" greatly encouraged belief in the atomic hypothesis of matter. Maxwell, Rudolph

Clausius (1822-1888) and others developed the ideas of this theory further, coming to the modern view of heat as a form of energy. Heat too, along with light, lost its status as an element. Perhaps Maxwell's most important contribution in understanding of the nature of matter was the view he took of atoms, hard and elastic but not "perfectly hard".

A new method of identifying elements developed from the flame test introduced long before by Boyle. In 1859, Robert Bunsen and G.R. Kirchhoff made a spectroscope and began to use it to examine spectral lines of elements. These lines can be seen when an element is vaporized and the vapor is heated to the point of emitting light. The lines appear when this light is broken into its various colors by a prism or other device. The light does not become a full rainbow but a series of sharply separated lines of different colors. The colors are a fingerprint of the element that emitted them. Through the application of this new technology, many more elements were discovered. The most notable of these was helium, first found in the Sun (hence, the name) and only later on Earth. By 1869, more than 60 elements were known.

The Periodic Table of the Elements dates from 1869 when Dmitri Mendeleyev (1834-1907) noticed that the chemical properties of the elements fall naturally into groups or families. In each family the properties change proportionally with the atomic weights. Therefore, Mendeleyev ordered the elements in a line by atomic weight and then cut the line periodically and started a new line underneath the first such that elements with similar properties are in the same column. This process created a chart with elements in columns of chemically similar elements and rows in which atomic weight increases from left to right. Most importantly, gaps appeared in the chart and the organization of the chart made it possible to predict many of the properties as well as the numbers of missing elements. The same idea occurred at almost the same time to John A.R. Newlands in England and Lothar Meyer. Newlands suggested the repetition was like octaves in music and he called it the "Law of Octaves". Unfortunately for Newlands, the repetitions do not always occur in sets of eight elements.

Finally, the failure of the Michelson-Morley experiment to detect changes in the speed of light eliminated aether both as a material and also as a real entity. Michelson, using the interferometer named for him, found no change in the speed of light from stars regardless of the direction of the star with respect to the motion of the Earth. The light from stars in the direction of the motion of the Earth had the same measured speed as light from stars in the opposite direction. Hence, the light was not moving in a stationary medium. It was not moving through a medium at all. There is no aether.

The Twentieth Century

New Phenomena

Just prior to the turn of the century, a flurry of major discoveries set the stage for dramatic changes in physics and chemistry in the early part of the new century. Wilhelm Roentgen (1845-1923) discovered X-rays in 1895 and, less than a year later, Henri Becquerel (1852-1923) discovered radioactivity (which we now recognize as radiation from the atomic nucleus).

Meanwhile, the English physicist J.J. Thomson (1856-1940) and Jean Perrin (1870-1942), a physical chemist at the University of Paris, had been studying cathode rays. In 1898 Thomson announced that the rays consist of minute, negatively charged particles with a mass at least a thousand times less than that of a hydrogen atom. He called them "electrons" (a word invented earlier by G.J. Stoney). Further, he suggested that atoms are made of rather spongy positively charged matter in which electrons were embedded. All this was greeted unenthusiastically until the American physicist Robert Millikan (1868-1953) measured the charge on the electron in his famous oil drop experiments of 1907-1909. Combining his results with Thomson's, Millikan was able to calculate the charge on the electron and the mass of the electron. The mass is almost 2000 times less than that of a hydrogen atom. Thomson was vindicated and interest was rekindled in the atomic hypothesis and in questions about the structure of the atom.

It was quickly recognized that X-rays (which come from the electrons outside the atomic nucleus) and radioactivity consist of very energetic rays and it was difficult to think of those large energies coming out of matter without connecting them to Prout's large forces for holding atoms together. H.G.J. Moseley (1887-1915) showed that X-rays could be used to place the elements in the Periodic Table. A detailed examination of the spread of X-rays from a particular element revealed that the frequencies of certain characteristic X-rays were directly proportional to the square of the atomic number of that element (starting with hydrogen as 1). Sadly, Moseley's research and life were cut short when the British sent him to the front lines in World War I.

Soon after the discovery of radioactivity, the Curie's began the painstaking chemical analyses that eventually led to the discovery of several radioactive elements, work Marie continued long after Pierre's untimely death (he was struck and killed by a truck in Paris). Marie Curie eventually received the Nobel Prize for this work.

By 1902, Ernest Rutherford (1871-1937) and his colleague Frederick Soddy (1877-1956) recognized that the alpha rays of radioactivity came from the breaking-up of an atom into a lighter atom plus the alpha particle.

Weighing Atoms

In 1827, the English botanist Robert Brown (1773-1858) called attention to the continual, erratic motions of small particles in solutions. Dust or pollen grains, under a microscope, are seen to execute perpetual zig-zags and jiggles. Lucretius (and very likely Epicurus before him) had pointed to the "multitude of tiny particles mingling in a multitude of ways" in a sunbeam and had noted:

...their dancing is an actual indication of underlying movements of matter that are hidden from our sight. There you will see many particles under the impact of invisible blows changing their course and driven back upon their tracks, this way and that, in all directions. You must understand that they derive this restlessness from atoms. It originates with the atoms, which move of themselves. [xvii]

However, those who followed Brown preferred to believe the motions were due to microcurrents in the air or water. They resisted an atomistic explanation. Ernst Mach and Wilhelm Ostwald were particularly influential opponents of atomism, emphasizing the hypothetical nature of atomistic theories and the hopelessness of ever observing an atom. Mach went as far as denying that the concept "atom" even belonged in physics since it did not correspond to an observable entity!

For the most part independently, three men renewed the consideration of Brownian motion around 1905. Albert Einstein published two papers on the subject in that year and has received most of the credit for explaining Brownian motion. However, Marian Ritter von Smolan-Smoluchowski (1872-1917), a Polish physicist working in Vienna and Jean Perrin came to the same conclusions as Einstein from much the same sort of reasoning. They all concluded that the zig-zags are not due to collisions of the moving particle with individual atoms or molecules but rather to collective or average collisions of about 10^{20} atoms or molecules per second with the particle. Furthermore, all three of them were able to show that the root-mean-square distance (a kind of average distance) a particle will move from its initial position in a given amount of time is proportional to the square root of that time. The particle will move twice as far in four seconds as in one second.

Additionally, they all realized that Avogadro's number could be found from data on Brownian motion. Einstein went on in that year to find a value for Avogadro's number from data on the density of sugar solutions. His first value was 2.1×10^{23} but, when improved data became available, he revised that to 4.1×10^{23}. When Einstein's students found an error in his equations, the value improved still further to 6.6×10^{23}. Eventually, Einstein found four distinct methods of finding Avogadro's number and his calculations gave values ranging from 6 to 9×10^{23}. The agreement of these values from highly diverse data and experimental situations did much to convince the doubters that atoms were real.

Avogadro's number is not an independent constant of nature. It is, in fact, the inverse of the mass of the hydrogen atom (in grams). If, as Einstein found, Avogadro's number is 6×10^{23} atoms per gram of hydrogen, then one hydrogen atom weighs 1 gram divided by 6×10^{23} or 1.67×10^{-24} grams. Hence, the determination of Avogadro's number gives us a good idea of the weight of an atom.

Perrin embarked on a program for getting Avogadro's number out of data on Brownian motion. His method required careful and numerous observations of individual particles under a microscope, recording their positions at equal time intervals. The results of thousands of observations were plotted and the graph analyzed, producing a value of Avogadro's number consistent with Einstein's values. In 1913, Perrin wrote a popular book, *Les Atomes*, in which he argued passionately and with all the considerable information available to him in favor of the atomic nature of matter. He regarded his own work as conclusive evidence of the atomic hypothesis. His contemporaries agreed; Perrin was given the 1926 Nobel Prize in Physics for his "work on the discontinuity of the structure of matter".

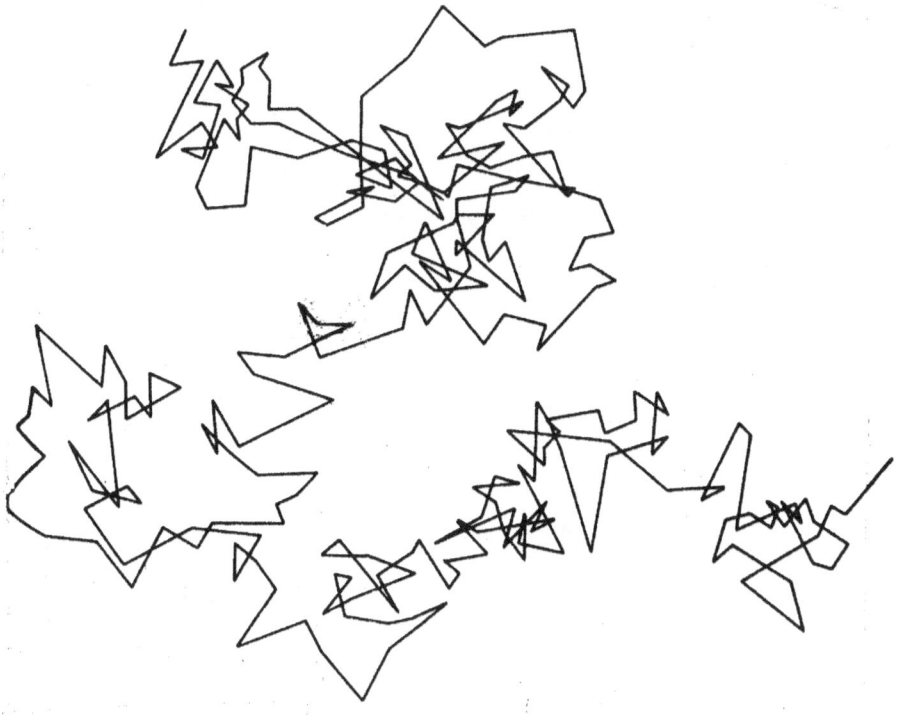

Figure 25. After Perrin's data of the zigzag path of a particle in Brownian motion.

What Are Atoms Like? The Quantum Idea

By the early part of this century, belief in the existence of atoms became a settled issue in physical science. But what atoms were like, what their structure was, was an open question. J.J. Thomson had challenged the hard, solid atoms of Newton and also Dalton's atoms whose only additional structural feature was the atmosphere of aether or caloric. The Thomson "plum pudding" model of negatively charged electrons embedded like raisins and currants in a spongy "cake" of positive charge was the early contender but it soon gave way to other ideas.

Two lines of development eventually coalesced to give our model of the atom; one was direct, the other indirect. The indirect approach grew from the discoveries of blackbody radiation and the photoelectric effect. Every object radiates electromagnetic radiation. The wavelength depends on the temperature but not on the material. Most objects are cool enough to radiate in the microwave or infrared region of the electromagnetic spectrum but at higher temperatures the radiation is visible light (e.g., red hot objects). This radiation, visible or not, is called blackbody radiation.

The curve produced when the intensity of blackbody radiation is plotted against the wavelength is peaked at a wavelength characteristic of the temperature of the emitting body. Nineteenth century physics had no explanation for the peaked curve. While this may not seem important to the non-physicist, it was catastrophic for physicists. Indeed, the best theory, the joint effort of Lord Rayleigh and Sir James Jeans, predicted a curve that steadfastly refused to peak but instead climbed continuously upward toward infinite intensity as the wavelength decreased. This behavior was labeled the "ultraviolet catastrophe".

In the last month of the 19th century, Max Planck pointed out that the Rayleigh-Jeans theory made the assumption that the radiation could be emitted at any wavelength with any energy. Though reasonable, the assumption was not necessary. Planck found he could get a peaked curve which fitted the data if he assumed instead that the energy of the oscillators was proportional to their frequency through a proportionality constant, h (Planck's constant) equal to 6.63×10^{-34} Js. That is, the energy, E, is related to the frequency, f, by $E = nhf$ where n is any positive integer. We call this limiting of a quantity to only certain values "quantization". For example, the channels you may select on a television are quantized while the volume selection is not quantized.

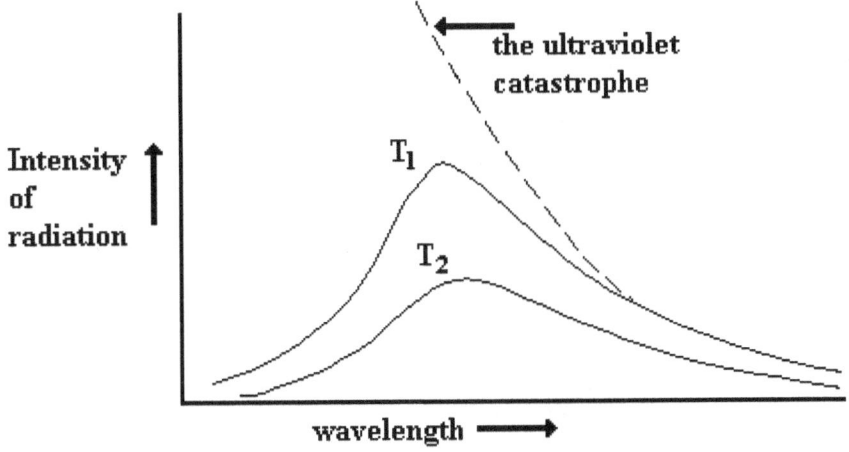

Figure 26. The Blackbody Radiation Curve with its puzzling peak.
The curves are for a body at different temperatures T_1 and T_2.

The photoelectric effect occurs when radiation (usually ultraviolet light), falling on a material, ejects electrons from the surface of the material. It is the phenomenon exploited in photocells. In 1905, Einstein picked up Planck's quantization idea to explain why the photoelectrons cannot be ejected from the surface of a given material unless the wavelength of the light is below a threshold value, regardless of the intensity of the light (Figure 27). Planck had quantized the energies of the light emitters in matter and, Einstein reasoned, the light, getting its energy from the emitters, must be quantized too. Ignoring the "fact" that light is a wave, Einstein showed that light arriving in small packets of energy would explain the threshold in the photoelectric effect. Assuming the surface keeps electrons bound so that a great deal of energy is needed for an electron to escape the surface, a packet of light with less than the necessary energy could not give an electron enough energy to be ejected from the surface. Einstein suggested that light, too, is quantized with energy equal to its frequency times Planck's constant so that E=hf.

Millikan measured the energies of ejected electrons in 1916 and showed that Einstein's predictions were correct. He also was able to evaluate Planck's constant from his data and it agreed with Planck's own value from blackbody radiation.

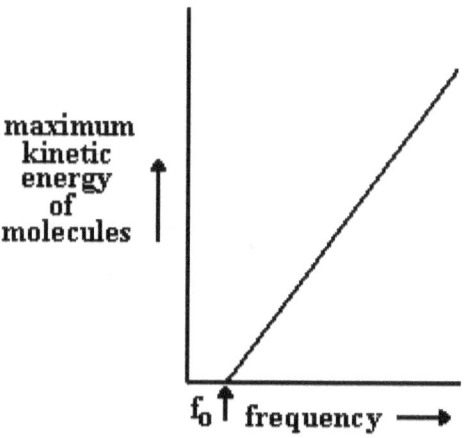

Figure 27. The linear plot of the photoelectric effect. f_0 is the threshhold frequency below which no electrons are emitted.

Thus the dual concept of light developed. Light behaves like a wave but, when small objects or small amounts of light are involved, light behaves like a particle. In order to agree with Einstein's theory of special relativity, any particle that travels at the speed of light can have no rest mass. The massless light particle is called a photon. It seemed to his contemporaries that Einstein was reinventing massless phlogiston as well as particles of light. His ideas were not well received initially for that reason. Nevertheless, quantization has become a major factor in the developments of the 20th century.

What Are Atoms Like? The Bohr Theory

Ernest Rutherford and his colleagues pursued the direct line of approach to the structure of the atom. Rutherford and Soddy identified the alpha particle as a helium atom stripped of its electrons. Rutherford quickly enlisted this particle as a probe of atomic structure. He assigned his post-doctoral assistant Johannes Geiger (1882-1945) and Ernest Marsden, a lowly undergraduate, the task of counting the number of alpha particles scattered at various angles after passing through a thin gold foil. Most of the alphas were undeflected but a few were scattered out of the main beam at all angles.

Rather as a clean-up exercise, they looked for particles reflected back from the foil without passing through. To their astonishment, they found particles bouncing back off the foil! A few particles were found to bounce straight back from the foil! Rutherford later remarked "it was almost as incredible as if you fired a fifteen inch shell at a piece of tissue paper and it came back and hit you".

Pondering the results for several years, Rutherford eventually showed that Geiger and Marsden's results implied the alpha particles were encountering an extremely dense, tiny, positively charged nucleus in the atoms of gold. Very few particles bounced back so the nucleus had to be very small and, hence, seldom struck. The nucleus had to be very dense or the heavy alpha particles would not bounce back. The nucleus had to be positively charged to make the atom electrically neutral. Accordingly, he advanced a planetary model of the atom in which electrons orbited the nucleus at distances much greater than the size of the nucleus. The model had one glaring defect. Maxwell's electromagnetic theory required any orbiting electrons to radiate energy in the form of light and spiral quickly into the nucleus. The model was extremely unstable and rapidly converted to a small, dense, neutral nucleus.

In 1910 Soddy realized that there was a simple solution to a problem developing from the discoveries of new radioactive elements. Elements with the same chemical properties but different weights were appearing. Soddy proposed that all the atoms of an element need not be identical. He called the different atoms of the same element "isotopes". In time it was realized that isotopes of one element have the same number of protons but different numbers of neutrons. James Chadwick verified the existence of the neutron in 1932. Now the peculiar values of atomic weight could be understood. Atomic weights of isotopes are very nearly integer multiples as Prout had believed. The atomic weight of any element is an average over the atomic weights of the naturally occurring isotopes of the element and need not be an integer and, indeed, is likely to be fractional.

Niels Bohr arrived in Rutherford's laboratory in 1912 as a postdoctoral student. He spent some time learning about the various problems being worked on before deciding on a particular project. The unsolved riddle of the unstable planetary model eventually captured his attention and for months he worked without success at making the model stable. He finally reached the point where he simply refused to permit the instability to exist. He realized that if light is quantized, as Einstein suggested, the solution might be that continuous radiation of light is not possible for electrons in orbit. If the electron could only radiate energy one chunk at a time, it would not spiral into the nucleus but move inward in jumps from one stable orbit to another.

Working out the mathematics of the orbits under an electrical attraction from the nucleus and assuming the electrons could only change from one orbit to another either by absorbing or emitting light of energy h times its frequency, Bohr found more information was needed. He was able to extract what he needed from his "correspondence principle". At great distances from the nucleus, the electron ought to obey the rules of the classical Maxwellian theory. Much closer to the nucleus, the new rules of quantization must take over. In between, the two sets of rules and behaviors should transform smoothly from one extreme to the other. The correspondence principle implied that orbital angular momentum is also quantized for the orbiting electron in units of h divided by 2p.

These ideas proved to be enough to calculate the radius of the hydrogen atom at 0.53 Angstroms (1 A = 10^{-10} m). The radius had already been measured as 0.50 A and Bohr knew he was on the right track. He found he could do something he had not expected to be able to do. The spectra of the elements are so complex that no one at the time expected soon to be able to predict them. Bohr found he could now calculate these wavelengths and his calculated values agreed exactly with the measured values!

The enormous success of the Bohr theory forced the rapid acceptance of the assumptions and ideas underlying the model. With modifications, they are the basis of our current thinking about atoms and atomic structure. The long years of indecision about the nature of matter ended in favor of the atomic hypothesis. The modifications have been in the direction of reducing our ability to envision the atom in terms of everyday experience. Louis de Broglie suggested, in his doctoral thesis of 1925, that the electron might be a particle with wave behavior just as light is a wave with particle behavior. He predicted a wavelength equal to h divided by the momentum (mv) of the electron. Davison and Germer in the United States quickly confirmed the prediction.

Erwin Schrodinger then founded wave (quantum) mechanics by finding a wave equation for describing the behavior of particles of atomic size. Schrodinger's equation has become our fundamental description of atomic processes. The modern atomic model is more an equation than a picture. The idea of mathematical order in the universe has become the idea of a mathematical universe. As Bohr remarked:

> To the physicists it will at first seem deplorable that in atomic problems we have apparently met with such a limitation of our usual means of visualization. This regret will, however, have to give way to thankfulness that mathematics in this field too, presents us with a tool to prepare the way for future progress.

Further Reading

The Architecture of Matter, Steven Toulmin and Jane Goodfield, Harper & Row, 1962.

Men Who Made a New Physics, B. L. Cline, Univ. of Chicago Press, 1987.

Niels Bohr, A Centenary Volume, ed. A.P. French and P.J. Kennedy, Harvard Univ. Press, 1985.

The Physical Sciences Since Antiquity, ed. Rom Harre, St Martin's Press, 1986.

"The Science of Matter", Robert P. Multhauf in *Science in the Middle Ages*, ed. David C. Lindberg, Univ. of Chicago Press, 1978.

The Atom in the History of Human Thought, Bernard Pullman, trans. Axel Reisinger, Oxford Univ. Press, 1998.

Chapter VI

Energy and Heat

The Energy Picture

Within the physical sciences, there are two main pictures of the causes of motion and change of motion. The force picture, which developed out of Newton's work, was the first of the two pictures to be elaborated in sufficient detail to be used for explaining and predicting physical events. As we have seen, this picture is powerful and compelling. Using it, however, requires knowledge and explicit inclusion of the directions of forces. In the second picture, where energy instead of force is cast in the lead role, explicit knowledge of directions of forces or even of motions is not required. Consequently, it is an easier picture to use in calculations.

Owing to its extreme subtlety, the energy picture developed very slowly. An important obstacle to its development was the allied problem of understanding the nature of heat. Heat had to be separated from fire and from matter generally before its connection to energy could be recognized. Also, Newton's inaccurate definition of force had to be corrected and the status, range of applicability and relative importance of the conservation of quantity of motion and of *vis viva* had to be understood. Thus, a great many lines of research, none aimed at the (at the time non-existent) concept of the conservation of energy, had to reach fruition before the outline of this great principle could be grasped. The history of this topic is, therefore, far more confused than that of any of our previous topics. By way of compensation, it is much shorter.

As has so often been the case, the story begins with the Greeks. It was Aristotle who coined the word *energeia* from *en ergon*, Greek for "at work". As we will see, this is an almost prescient invention since energy, in the physicists sense, is now closely tied to the idea of mechanical work. The word was appropriated by the practical Romans, who knew a useful word when they found one, and has been taken over into English from the Latin form *energia*. However, as with other words such as "power" and "force", "energy" was long in common use before it came to have a technical meaning. For example, at the end of Newton's lifetime we find Alexander Pope praising John Dryden's poetry for its "…varying verse, the full resounding line, The long majestic march, and energy divine"[xviii]. It was much later that the word was explicitly defined in physics.

The earliest usage was the suggestion by Thomas Young, in 1807, that the term *vis viva* should be replaced with the word "energy". The suggestion was ignored until about 1851 when Lord Kelvin remarked that heat gives rise to mechanical work and "intrinsic energy." This amounts to an early statement of the First Law of Thermodynamics that is acceptable today although we would say "internal" rather than "intrinsic" energy. Then in 1855, William Rankine (1820-1872) began to propound "the science of energetics" (a form of the principle of energy conservation). He used the term "potential energy" to replace many of the uses of "latent heat" and called energy of motion and heat by the general name "actual energy".

Principles of Minima

Some of the earliest thinking which eventually contributed to the ideas of energy and energy conservation came from optics. Hero of Alexandria gave a proof that light takes the minimum length path traveling from one point to another. In his ***Catopterics***, he used this principle to explain why light reflects from a surface at the same angle at which it strikes the surface. The idea lay unused until Pierre Fermat (1608-1665) altered it subtly, making it into a principle of least time rather than least distance. The change was significant because a principle of least distance cannot be applied to refraction where the light is made to travel through different materials on its path. Least distance applies easily to cases of reflection such as a light beam bouncing off a sheet of glass. But it is useless for refraction such as the passing of light through the lenses of a pair of eyeglasses. Fermat was able to derive Snell's law of refraction as well as the law of reflection from this principle of least time. Johann Bernoulli had, somewhat earlier, obtained the same results with a similar line of reasoning.

Pierre L.M. Maupertuis (1698-1759) expanded Fermat's principle into "le principle de la moindre quantite d'action" (the principle of least action). He defined action as the product of the mass, velocity and distance of the motion of a body. Today, action is the product of energy and time and has the same units as the original "action." Planck's constant, h, is the minimum quantity of action. Unfortunately, Maupertuis was inconsistent in the way he calculated these quantities so the examples he gave showing the principle in use were more confusing than enlightening. Léonard Euler refined Maupertuis' ideas in 1744 but restricted their use to situations where the velocity of the object depends only on the position of the object (we would say where the object is acted on by a conservative force).

Sixteen years later, Joseph-Louis Lagrange (1736-1813) showed that the principle of least action is mathematically equivalent to Newton's laws of motion. Lagrange's equations of motion, however, are more easily expressed in terms of what we now call energy. Thus, a second picture of the causes of change of motion arose in physics. Newton's picture represents forces as the causes of change of motion. Lagrange's energy picture represents motion as controlled by energies. Another minimal principle closely but not obviously related to Lagrange's work was established by Jean d'Alembert. He found that Newton's equations of motion for an object could be rewritten to minimize the sum of a series of terms. Each term is the product of a force on the object and the distance the force moves the object. D'Alembert's Principle converts rather easily into Lagrange's results but the quantity emphasized is different. In modern terms, d'Alembert's Principle deals with mechanical work while Lagrange's equations are expressed in terms of potential and kinetic energies.

Conserved Quantities

Another series of developments that contributed to the final emergence of the concepts of energy and energy conservation were the ideas about conserved quantities generally. There is a close connection between conserved quantities and minimal quantities. Mass conservation is an especially ancient idea albeit an inexact one initially. Lucretius and the Greek atomists emphatically saw atoms as indestructible. That, after all, was part of their intended function as changeless entities in a world of changing things. The atomists held that the number of atoms in the universe is constant or conserved. They did not think of mass in a modern sense, if only because they could not separate the mass of an object (its quantity of matter) from its weight (the force of attraction exerted on the mass by the Earth). Had they explicitly tied mass to atoms, they would have had a thoroughgoing doctrine of the conservation of mass. No other Greek thinkers developed anything like the doctrine.

Not until the 17th century in England and France when men like Boyle and Rey began using mass (weight) as a quantity of importance in the quantification of chemical processes did mass conservation again become a seriously considered possibility. With Newton's work came a precise definition of mass and the establishment of a clear distinction between mass and weight. Lavoisier's work further demonstrated the power of the idea for clarifying details of chemical reactions as did Dalton's work a few years later. Mass conservation occupied a secure position among the physical principles by the beginning of the 18th century. Fortunately, Einstein's work, implying the interchangeability of mass and energy, came too late to muddle the developing understanding of energy conservation.

Other important conserved quantities were known by 1800 and have much shorter histories. We have seen how the conservation of *vis viva* and of momentum were quickly recognized through the efforts of Descartes, Huygens and Newton. Momentum is absolutely conserved but *vis viva* is not. Huygens found he had to require conservation of *vis viva* to explain certain collisions. However, *vis viva* is not conserved in all types of collisions. A tremendous amount of time and effort had to be expended to determine the exact limits of these conservation rules.

The idea of heat conservation has a more complex history. Galileo's invention of the thermometer in 1595 and subsequent improvements of the instrument by others made accurate, objective temperature measurements possible. By 1701 we find Isaac Newton reporting that if the temperature registered by the thermometer "in packed snow, while it is melting" is taken as zero, then "the heat at which the water boils violently" is 34 "parts." He noted the temperature remained essentially unchanged throughout the boiling process, saying "water begins to boil at 33 parts and scarcely reaches a heat of 34 1/2 parts when boiling." A year later, Guillaume Amontons (1663-1705) observed that boiling "water cannot acquire a greater degree of heat however long it is on the fire..."

The Scottish physician, Joseph Black, carried out a most thorough thermometry. Although he published little in his lifetime, his work made itself felt through his lectures in Edinburgh, the notes of which were published posthumously in 1803. He was the first to see that heat and temperature must be distinguished, warning against "confounding the quantity of heat in different bodies with its general strength or intensity [temperature]…" Black was also the first to realize that different materials have different capacities for receiving or giving off heat. As he pointed out, this heat capacity of a material has no simple connection to the density of the material. As an example, he compared water and mercury. Mercury has a density 13.6 times that of water but its heat capacity is much less per gram. Developing the idea of heat capacity, he came to understand that, through a sharing of the heat, bodies in thermal contact come to have the same temperature (but not the same heat content) at what he called the "equilibrium of heat". Heat flows from hotter objects to colder objects until both are at the same temperature.

Black also examined more closely than Newton or Amontons the lack of change of temperature as water boils or freezes. He coined the phrase "latent heat" to explain how, at these points, the addition of heat does not change the temperature but instead changes the water into steam or ice into water. His idea was that the heat fluid of latent heat combined almost chemically with the water. Finally, he noted, "when the vapour of water is condensed into a liquid, the very same great quantity of heat comes out of it into the colder matter by which it is condensed." So heat was not lost in the process but conserved.

From Black's work arose a vision of a heat fluid, named caloric by Lavoisier, which could neither be created nor destroyed. Because heating a material did not seem to change the weight of the material, the heat fluid was thought to be weightless. Furthermore, to explain expansion of heated materials, it seemed necessary to view the particles of caloric as attracted to matter but repelled by each other. Thus, when heat was added to a material, the particles of heat distributed themselves rapidly through the material and formed shells or atmospheres around the particles of material (figure 24). Because caloric was conserved, and because conserved quantities like mass are associated with matter, Lavoisier and others regarded caloric as a material substance.

The man whose work went far towards undermining the idea of heat as matter was originally an American. Benjamin Thompson (1753-1814) served briefly in the American Revolutionary army but out of pique over a slight or possibly out of real Tory sentiments, he went AWOL. He fled to England and enlisted the support of Lord George Sackville in some scientific projects he hoped to pursue. Eventually, after numerous adventures, he became the Bavarian minister of war and was given the title "Count Rumford" to go with an English knighthood already received. The name of Rumford he himself chose in honor of his hometown, Rumford, Hew Hampshire. He married the widow of Lavoisier but the marriage seems to have been one of his less successful adventures.

In his capacity as minister of war, Rumford had access to some of the best equipment in Europe for accurately weighing large objects. He went to work trying to find the weight of the latent heat of fusion of ice (the amount of heat needed to melt a gram of ice). He was unable to find any weight and he soon abandoned the effort, remarking that he did not expect the weight would ever be found. The weightlessness of caloric did not greatly trouble him or his contemporaries, however odd it may seem to us.

Rumford next discovered the much more damaging fact that caloric was not conserved. As minister of war, he was responsible for the boring of cannon. This task, carried out by teams of horses, involved drilling the large brass castings from which the cannon were made. The horses walked around a large circle harnessed to a vertical auger made to turn with the circling horses. The rotating auger cut into a brass casting as it turned, forming the bore of the cannon. In the process, large quantities of hot brass chips were generated and the castings themselves became quite hot. He remarked, "the more I meditated on these phaenomena, the more they appeared to me curious and interesting." The explanation of "these phaenomena" based on the caloric theory was that caloric was squeezed out of the chips and into the body of the casting. Rumford thought it odd that the chips should then be hotter than the cannon as if their heat capacity was different from that of the bulk metal. Repeated measurements of the heat capacities by Rumford revealed no such difference.

In 1798, Rumford carried out an ingenious and carefully executed series of experiments with a very dull auger. He found the temperature of the cannon was raised some 70° F while only a small amount of brass dust was drilled out of the cannon body. Heat seemed to be created inexhaustibly. He concluded that caloric could not be a "material substance" and that heat must be related to motion.

The motion of the horses was being communicated to the brass casting and was appearing as heat in the casting. Rumford therefore began to promulgate the "very old doctrines which rest on the supposition that heat is nothing but a vibratory motion taking place among the particles of the body."

This "very old doctrine" was, of course, an *atomistic* doctrine and as such was inconsistent with the caloric theory that was still essentially a *continuum* theory of matter.

Although Rumford's work considerably weakened the caloric theory, the theory was not overthrown. As we have seen before, a defective theory will not be abandoned unless and until a better theory is available. The new theory must not have as many deficiencies as the old one and it must explain everything the old theory explained. Preferably the new theory should explain more than the old and should also improve the accuracy of predictions. Rumford had shown that caloric could not be conserved but that did not mean it did not exist. The "old doctrines," he favored, connected to the dubious atomic theory, did not yet constitute a well-worked out theory. It was possible to doubt their explanatory powers; the situation did not change until a half century later with the work of Joule and Kelvin.

Technological Developments

Additional reasons for seeing a connection between heat and motion came from a new technology. In 1712, Thomas Newcomen invented the first functional steam engine. It was used for pumping water out of coalmines but was ponderous and not especially efficient. James Watt patented the first efficient steam engine in 1769 and it quickly supplanted the Newcomen engine. Watt gave his engines away! The recipient, in lieu of purchasing the engine, agreed merely to pay Watt the savings in fuel costs over the first three years of use. Eventually the steam engine was adapted to hauling coal and then to carrying passengers.

The most elementary thinking about this technology necessarily suggested a connection between heat and work. Heat is supplied to the engine and the engine responds by doing mechanical work, putting in motion water or the engine itself. The engine turns heat into motion.

Considerations such as these attracted the critical attention of a young French army captain of engineers, Sadi Carnot (1796-1832). He was the son of Lazare Carnot, an able mathematician, military engineer, general and hero of the victory of the French Republic. His father was a major figure in the development of a theoretical understanding of mechanical engines. Carnot, the son, extended his father's work to steam engines. Sadi was educated under his father at the Ecole Polytecnique, probably the best technical education then available. Consequently, he was better prepared to pursue a theoretical approach to the operations of the steam engine than anyone in the British Isles where steam technology was well developed but without a solid theoretical underpinning.

Carnot began by noting the cyclic pattern of the operation of the engine (the piston goes back and forth). The net effect of each cycle, he saw, was to "transfer [caloric] from a hotter to a colder body" and not to cause a "consumption of caloric" as was usually assumed. The steam engine (Carnot preferred to call it a heat engine) operates by making motion from this transfer or flow of heat. Idealizing the cycle and the process, he imagined an abstract engine operating by having the flow of caloric channeled through it from a hot heat reservoir to a cool heat reservoir.

The engine extracted motive power from the flow of caloric. Carnot was able to prove that "the motive power of heat is...determined solely by the temperatures of the two bodies between which...the transfer of caloric occurs." Furthermore, Carnot was able to calculate the maximum efficiency of a heat engine. If Q_1 is the heat extracted from the hot reservoir and passed to the heat engine and if Q_2 flows to the cool reservoir from the engine while the engine extracts W useful work from the heat flow, then the efficiency of the engine is just W/Q_1. Assuming the conservation of caloric, $W = Q_1 - Q_2$ so the efficiency can also be written $(Q_1 - Q_2)/Q_1$. Carnot then invented a process, the Carnot cycle, which he was able to prove must give a maximum efficiency. The maximum efficiency is $(T_1 - T_2)/T_1$ where the T's are the absolute temperatures of the respective hot and cold reservoirs. It is called the Carnot efficiency.

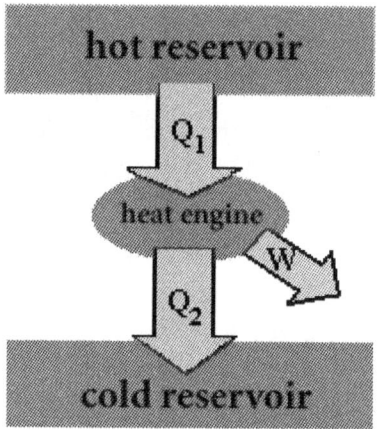

Figure 28. Carnot's model of the workings of a heat engine.

Carnot's value is in fact an overestimate. Friction is present in all machines, reducing their efficiency. Also, his model is too simple. For heat to flow from the high temperature reservoir to the heat engine, the heat engine must be cooler than the reservoir. Likewise, heat will not flow from the heat engine to the low temperature reservoir unless the engine is warmer than the reservoir. Thus, we must insist that the heat engine have a temperature in between those of the reservoirs. Taking this factor into account gives a lower efficiency than the Carnot efficiency.

Despite the limitations, a major question of the machine age was answered. Finally it was possible to consider how much work an engine could be induced to produce. And the answer is—not as much as one would like. In fact, the useful work output, W, will never be as great as the heat input, Q_1. In this research Carnot became the first person to use the first two laws of thermodynamics, though, of course, he had no idea he was doing so and his efforts in that direction went unappreciated for some 25 years. He himself died of cholera at age 36 so he had no opportunity to follow up this seminal work.

Heat as Molecular Motion

Although various people at different times had suggested heat might be somehow related to motions of the invisible particles of matter, not until Daniel Bernoulli (1700-1782) published his book *Hydrodynamica* in 1738 was the vague idea refined into a functional theory. Bernoulli proposed, as a model for "air" (by which he meant gases generally), that we imagine the gas consists of "very minute corpuscles", spherical in shape, "driven hither and thither with a very rapid motion." Gas pressure was the result of multiple collisions of these corpuscles with the walls of the container. The model rather easily led to a derivation of Boyle's Law. Bernoulli then correctly remarked that experimental verification of the law required constant temperature.

He explicitly connected the "very rapid motion" of the corpuscles with heat, saying "heat may be considered as an increasing internal motion of the particles." Thus, he single-handedly laid the foundations of the kinetic theory of gases in which heat is associated with the random, rapid motions of molecules.

Bernoulli's work was generations ahead of its time. The phlogiston theory still dominated thinking about heat in 1738 and his results were regarded as anomalies rather than profound insights. The demise of the phlogiston theory produced no change in this attitude, possibly because that theory was so rapidly replaced by another fluid theory of heat, the caloric theory.

By the end of that century and into the early part of the nineteenth century, new information and new attitudes began to appear. Rumford's experiments led him to advocate a kinetic theory of heat (that is, the idea of heat as a random motion of the constituent "atoms" of matter) and a relatively unknown experiment caused Joseph Louis Gay-Lussac to doubt the caloric theory. In 1807, Gay-Lussac allowed a gas to expand into a vacuum. The temperature of the gas remained what it had been before the expansion. Caloric theory required a temperature drop in such an expansion because, in the theory, temperature was proportional to the concentration of caloric and the concentration had to drop with an increased volume.

Despite the counter-indications, the caloric theory retained its influence until the middle of the nineteenth century. Dalton thought of his atoms as surrounded by atmospheres of caloric. Carnot's theoretical investigations of the heat engine were based on the caloric theory. It was a useful if not perfect theory and the replacement option, atomic theory, was not an obvious improvement. The power of the atomic theory was not yet visible.

Bernoulli's ideas finally received the attention they deserved when James Joule (1818-1889) and William Thomson, Lord Kelvin (1824-1907) began their occasional collaborative work. In 1845, Joule repeated Gay-Lussac's experiment obtaining the same result. Reporting this he commented, "no change of temperature occurs when air is allowed to expand in such a manner as not to develop mechanical power." He had noticed that free expansion (into a vacuum) produces no temperature change but expansion against a resistance, e.g., where the expansion moves a piston against air pressure, the temperature decreases. The temperature decrease is used to liquefy gases. Another important connection between heat and mechanical work had been made.

Joule continued his efforts. In 1851 he published the first calculation of the speed of a molecule. Taking a cubic foot of hydrogen at 60° F, he found that, regardless of the number of particles in this amount of hydrogen, their speed must be 6225 ft/s! He was using Bernoulli's basic model and, not surprisingly, found Boyle's Law as a by-product of the calculation. He went on to observe that since, in the model, "the pressure is proportional to the squares of the velocity of the particles, in other words, to their *vis viva*, it follows that the absolute temperature, pressure and *vis viva* are proportional to one another..." That pressure is proportional to temperature was known (Charles' Law); now Joule had shown the kinetic theory accounted for that law. Even more importantly, Joule's remark points out that the temperature, Black's measure of the intensity of heat, is in fact a measure of the product mv^2, the *vis viva*, of the average molecule.

Others now joined the effort. Maxwell showed that, in the kinetic theory, molecules do not all have the same speed but a distribution of speeds. The speed Joule had found is that of the average hydrogen molecule although at any given time it might be that no individual molecule has exactly the average speed. Temperature is proportional, then, to the average *vis viva* of all the molecules. The theory itself became more sophisticated with the spherical molecules replaced by non-spherical molecules of various shapes depending on the particular gas of interest and its molecular structure. The kinetic theory flourished and finally supplanted the fluid theories of heat.

The Mechanical Equivalent of Heat

All these historic trends which we have traced coalesced in the mid-nineteenth century around a number of intertwined understandings. One of these, the sense of a connection between heat used and mechanical work done by an engine was given its modern clarity and form by James Joule, a pupil of John Dalton who, like Dalton, was a confirmed Newtonian. From the very outset of his scientific work one can discern a fixed and clear conviction, amounting almost to a Kepler-like obsession, that "the grand agents of nature are, by the Creator's fiat, indestructible; and that wherever mechanical force is expended, an exact equivalent of heat is always obtained." From 1843, Joule measured this mechanical equivalent of heat in a wide variety of types of experiments. He measured the heat produced by an electrical current and compared it with the mechanical work done to produce the current in a "magneto-electrical machine." He compared the work needed to compress air with the heating of the air produced by the compression. He used friction to generate heat, churning water with paddles (Figure 29) and compared the heating to the work done in turning the paddles. He even went over Rumford's data on the boring of cannon and generated a number from that source. Lord Kelvin reported finding him, early one morning, trying to measure the temperature increase of water that had gone over a falls. Kelvin was impressed with Joule's persistence, all the more so because he knew very well that, at the time, Joule was on his honeymoon. Joule himself commented in a paper that he calculated the water at the bottom of Niagara Falls should be raised "about one fifth of a degree by its fall of 160 feet."

To an extent, Joule was hampered by the lack of standard units and measurements. Watt, for example, rated his engines in horsepower, the power a single horse could expect to continuously exert. Helpfully, Watt estimated this power, in English units, as 33,000-ft lbs per minute. That is, a horse could continuously lift a one pound weight 33,000 feet in one minute or 550 feet per second. Therefore, Joule rated all mechanical work in terms of raising weights. Work was rated in pounds lifted times the distance of the lift. Heat, on the other hand, was rated by its ability to heat water. One calorie (1 cal.) of heat was the amount needed to raise one gram of water one degree Celsius (Centigrade).

Comfortable with English units, Joule usually rated heat in units of the amount that raised one pound of water one degree Fahrenheit (252 calories). Stated in metric units, Joule found that one calorie of heat is always equivalent to the work needed to raise 4.186 Newtons of weight one meter. The Newton is the unit of force and 1 Newton x 1 meter is now called 1 Joule (1 Nm = 1 J). Hence, the mechanical equivalent of heat can be written

$$1 \text{ cal} = 4.186 \text{ J}$$

The statement compresses and summarizes 40 years of the life and work of one man, James Joule.

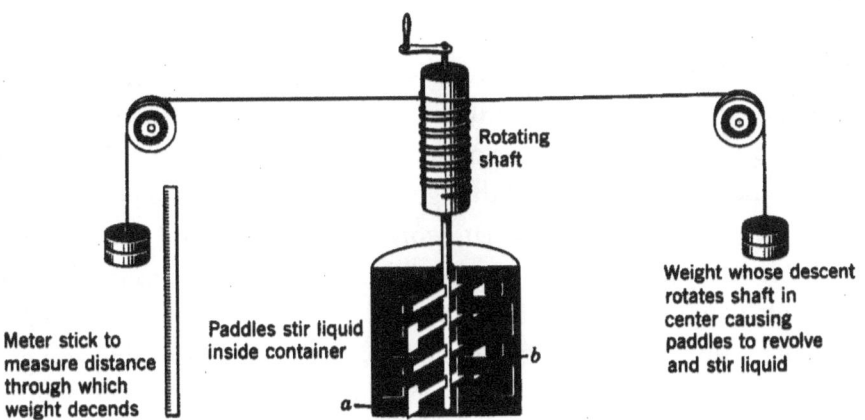

Figure 29. James Joule's paddle wheel experiment for heating water.

Please note: the idea here is that a certain amount of mechanical work converts to a certain amount of heat. However, it is not correct to reverse the statement and assume a certain amount of heat converts to an exact amount of mechanical work. In this difference lies the tale of the second law of thermodynamics. That is, 4.186 J of mechanical work can be turned into 1 calorie of heat but the reverse is not true. 1 calorie of heat cannot be turned into 4.186 J of mechanical work. The mechanical work available from 1 cal. must be less than 4.186 J.

The Conservation of Energy

Joule's idea of the mechanical equivalent of heat was by no means developed in a vacuum. The German physician Julius Robert Mayer (1814-1878) published a little noticed statement of the idea in 1842 in an obscure journal and, on the basis of the date, he and Joule eventually became embroiled in a priority dispute. Hermann Helmholtz (1821-1894) proposed the same idea in an 1847 paper presented to the Physical Society of Berlin.

Mayer had attempted to publish his paper in physics journals but it was refused. As a relative novice in physics, Mayer expressed his ideas in unconventional and somewhat confusing terms. He used the word "kraft" when work or energy was meant (although in fairness in must be noted that Helmholtz, too, used this German word for "force" in the same way) and he called forces "indestructible, convertible imponderable objects." However, he plainly connected heat, motion, and work. He even calculated the mechanical equivalent of heat in the paper, saying, "the warming of a given weight of water from 0° to 1° C. corresponds to the fall of an equal weight from the height of about 365 meters." This result is equivalent to 1 cal = 3.577 J, a value comparable to some of Joule's results.

Mayer's interest went beyond that of Joule in that Mayer, from the very beginnings of his work, obviously aimed at a broader equivalence. Whereas Joule focused especially on heat and work, Mayer included chemical and biological systems and effects within the scope of his principle as well as electrical and motional (kinetic) energies. Mayer's work long remained obscure, however, because of its peculiar character. Though Mayer is generally accorded today the distinction of priority, historically, it was Joule whose work most influenced the developing consensus about energy.

Helmholtz' paper was a major, mainstream advance. Joule's forte was experiment. It was Helmholtz, originally a physician like Mayer, who developed the theoretical side of the question. He showed that Huygen's idea that *vis viva* was conserved was true only for certain situations where the forces involved originated from "material points" and act "in the direction of the lines which unite [the points], and the intensity depends only upon the distance." Furthermore, consistent with ideas developed by the French physics and engineering community, he showed that work and *vis viva* were together conserved subject to the revision of the formula for *vis viva* from mv^2 to $\frac{1}{2}mv^2$. Finally, in the same paper he went on to fit heat, work, the redefined *vis viva,* and electromagnetic energies all into a sweeping conservation law for energies.

Even with Helmholtz' work, full clarity of vision on the subject of energy had not been reached. Greater clarity was introduced when, about 1855, Rankine began to promote the energy picture of physical events under the appellation "the science of energetics". A passionate coiner of new words, Rankine distinguished stored energy from the work done to create it, calling the former *potential energy* and the later simply *mechanical work.*

Stored energy such as that in a compressed spring or a hanging weight is created when a force is applied against the spring or against gravity and produces a change in the position of the end of the spring or of the height of the weight. Thus, potential energy is energy of position. The production of the change in position is mechanical work and, once it has been done, it becomes stored energy with a potential for creating motion of the spring or the weight. Thus, work is done and potential energy is created. On release, the potential energy transforms into kinetic energy, energy of motion. If the motion is absorbed, for example, when the weight lands and stops moving, the kinetic energy has been transformed into heat in the weight and the floor and sound. *This perpetual interchange of energy from one form to another is what is meant by the conservation of energy.* The total energy, distributed among all the forms, is constant. Only the relative distribution is changed. It is like an account book where the numbers in the different columns may change but the total of all columns, must remain the same.

The minimal principles, which to this time had seemed to be a set of inexplicable and independent quasi-metaphysical constraints on physical systems and events, could now be seen in a new light. They were now recognizable as corollaries, even special statements, of the conservation principle or as statements of the additional requirement that a system, to be stable, must be at its lowest possible total energy.

The Laws of Thermodynamics

Much of the initial thinking leading to the conservation principle grew out of practical considerations of practical objects: steam engines. The conservation of energy was not the only insight to come from the study of these engines. Among the questions raised by the mere existence of such engines were questions of how much one could hope to get from an engine. Is it possible, for example, to make a machine that, over one operational cycle, will put out more energy as work than it takes in as heat? Such a (non-existent) machine is known as a *perpetual motion machine of the first kind.* If we lower our sights a bit, perhaps we might hope for a machine capable of taking in heat from a high temperature reservoir and turning it all into useful work, dumping no heat to the low temperature reservoir. Obviously, the lower reservoir is then superfluous and can be eliminated. Such an also non-existent machine is known as a *perpetual motion machine of the second kind.*

Both Rudolph Clausius (1822-88) and Lord Kelvin recognized that a perpetual motion machine of the first kind would violate the conservation of energy principle. Reformulating the conservation principle to fit the situation, both stated in various forms what is now called the first law of thermodynamics: in a given time period, heat flow into a system equals the work the system does on its surroundings plus the increase in the internal energy of the system. A perpetual motion machine of the first kind would obviously violate this law because it would take in too little heat to supply the work done. It would create energy *ex nihilo*!

As to the less ambitious hope of creating a perpetual motion machine of the second kind, Kelvin and Clausius saw different but, as Kelvin proved, equivalent problems. Kelvin showed that the basis of Clausius' views on the subject was this principle (following Kelvin's wording): "It is impossible for a self-acting machine, unaided by any external agency, to convey heat from one body to another at a higher temperature." By "self-acting", Kelvin meant something running on its own. Kelvin's own statement was, "It is impossible, by means of inanimate material agency, to derive a mechanical effect from any portion of matter by cooling it below the temperature of the coldest of the surrounding objects." Perpetual motion machines of the second kind necessarily violate both of these principles.

Paraphrasing and simplifying these conclusions, we might say the first law is that energy is conserved-including heat. The second law would similarly become only part, not all of a given quantity of heat can be turned into useful work. The ultimate in simplifying these laws was reached when someone, a gambler most likely, paraphrased them: the first is that you can only break even; the second is that you can't even do that!

Clausius, in particular, was not entirely comfortable with such expressions of the second law. The first law, as an expression of a conservation principle, was satisfying. Seeing events as constrained by a limited supply of energy made good, intuitive sense for him, as it does for most people. Clausius wanted a way of expressing the second law as a simple constraint on something. But what quantity could the something be since the law denies 100% conversion of heat into work but does not have any other absolute limit?

A number of years of work and thought were required but eventually Clausius found a new quantity, even more abstract than energy, which is constrained in a simple way by the second law. It is a quantity that measures the disorder of the system and Clausius gave it the name entropy (transformation). He remarked:

> I have intentionally formed the word entropy so as to be as similar as possible to the word energy, since both these quantities…are so nearly related to each other in their physical significance that a certain similarity in their name seemed to me advantageous.[xix]

By naming it "transformation", Clausius indicated that it was to be a measure of how much heat could be transformed into work. The constraint on entropy tells one what fraction of the optimum transformation has been actualized. Clausius now could state the second law: "The entropy in the universe tends toward a maximum."

Clausius himself never seems to have appreciated the statistical nature of entropy. Kelvin and Maxwell, and then most particularly Ludwig Boltzmann (1844-1906) and the American, Josiah Willard Gibbs (1839-1903), developed statistical mechanics, which was in contrast with the more axiomatic and phenomenological structure of thermodynamics proper. Within a statistical structure, entropy is proportional to the number of possible states of the system, and the tendency for entropy to maximize is the result of the tendency for a system to gravitate towards its most probable states. The most probable states are also the most disordered (as any one knows who has not picked up his room in a week). Hence, Clausius' earlier association of entropy with disorder is apt.

The second law is unique among physical laws because it requires a direction for time. All other physical laws would work as well if time were to reverse and run into the past rather than toward the future. But if time were to reverse, we would see chaos tending toward order, systems drifting away from their most likely states and entropy decreasing rather than increasing. The law has been called "time's arrow" because of this directional character. In fact, as Clausius himself realized, the law implies that the universe will eventually run down to a state with all work and energy turned into heat. This "heat death" of the universe is, as one might suspect, a very long time away. Other events might intervene to terminate the universe well before its heat death.

The Twentieth Century

Since kinetic energy (Helmholtz' redefined *vis viva, ½mv²*) depends on the mass of an object, Einstein's paper concluding that mass must depend on speed forced a revision in thinking about kinetic energy and mass. In the paper Einstein showed that $E = mc^2$ is the kinetic energy plus the "rest mass energy" of the object and c is the speed of light. Since c^2 is so large $(9 \times 10^{16} \ m^2/s^2)$, the energy of even a small amount of mass is quite large. Prophetically, Einstein commented in the paper that it might be possible to confirm this relationship by examining the energy output of "radium salts". It is precisely in the area of radioactivity and nuclear energy that the relationship has proved to be most useful.

Thus, mass too has become a form of energy. The conservation of energy accountants must add another column to their books. Along with the columns for work and kinetic energy and so on, we must now have also a column for mass changes. However, mass conservation is no longer an exact or strict rule. Mass can disappear, turning into energy (usually in the form of a gamma ray). Mass can also suddenly appear. This commonly occurs in the upper atmosphere of our planet where a high-energy electromagnetic, cosmic ray photon disappears and a pair of particles, an electron and its antimatter partner, the positron, appear.

Another insight of the twentieth century was that all conserved quantities are related to symmetries of the universe and the physical laws representing the universe. Emmy Noether (1882-1935), often rated as the greatest female mathematician in history, was the first to point this out. Energy is associated with the fact that the universe looks the same from any place (translational symmetry).

Further Reading

The Second Law, P.W. Atkins, Scientific American Books, 1984.

Chapter VII

Electricity, Magnetism and Light

The Classical Period

The story we will trace in this chapter is even more tangled than that of the previous chapter and it is also historically longer. As with the story of heat and energy, phenomena that at first seem wholly unrelated we will, in the end, discover to be interconnected in profound and subtle ways.

Three different threads make up the tapestry of this story: electrical phenomena, magnetic phenomena and light. Close examination will reveal that each of these three threads is itself constructed of yet finer threads, which also seem unrelated initially. Nonetheless, they have been woven into a single, astonishing fabric. And it is not certain the process is finished.

The beginnings of the story are uncertain; no specific names are attached to it. Clearly, Greeks of the classical period were familiar with a number of electrical phenomena. Lightning and the auroras were known but not known to be related. The Greeks also knew that amber, when rubbed with cloth or fur, becomes capable of attracting light objects such as bits of straw or pieces chaff. This observation eventually gave us our English word "electricity"; the Greek word for amber is *elektron*.

Very early in the Iron Age, the ability of lodestone to attract pieces of iron (as well as other pieces of itself) was noticed. Lucretius even tells of seeing rings of gold-plated iron as well as iron filings move ***away*** from the stone. He did not know about magnetic repulsion between like poles and hence did not understand what he was seeing. Lodestone is the magnetized form of the iron mineral, magnetite. Pliny the Elder tells of a shepherd named Magnes whose iron tipped staff and iron nailed shoes stuck in a field where lodestone was abundant. Such fables aside, our word "magnet" comes from Greek name for the lodestone *magnes lithos* (stone of Magnesia), named for a region in east-central Greece where a number of important metals were mined. The attraction of the lodestone for iron is very similar to the attraction of small objects to rubbed amber. Though the similarity probably was noticed, the connection came to nothing since neither of the two phenomena was as yet understood.

The third thread, light, was obviously known to the classical Greeks but there was no hint of a connection between light and electricity and magnetism until the middle of the nineteenth century. Consequently, ideas of light developed quite apart from thinking about electricity and magnetism.

The story of our understanding of light illustrates how the development of science depends on the questions one asks and the presuppositions one brings to the endeavor. Our story will be centuries along its path before anyone even learns to consistently distinguish between light and sight, illumination and seeing. Light itself could not be studied without the physiology and psychology of vision intruding because all the information used in a study of light came through seeing and might, therefore, depend on the nature of seeing. Hence, more than any other area in the physical sciences, the study of light infringes on philosophical questions of how we know what we think we know. The study of light can become the study of epistemology or even of metaphysics. Indeed, it is hardly an exaggeration to say that in all religious traditions, whether pagan, monotheistic or animistic, light was regarded with awe and was frequently associated with the Divine or Sacred. It first began to lose its elevated status with the Greek philosophers.

Plato began this process of demotion. In a paragraph of the *Timaeus*, he advanced a theory which was to dominate one and a half millennia of thinking about light and seeing. Parenthetically, our word "theory" comes from the Greek *theoria* that means "to see" or "to behold". Light for Plato was a sort of animated fluid.

Plato thought of seeing as a human activity rather than a passive reception of light. He thought of the eye as a lamp that emits fire. Hence, the view he developed is called an "extramission" theory of seeing because, in it, the light comes *out of* the eye. During the day, the light of day combines with this visual stream from the eye (likes attract) "forming a single body in the line of sight". Plato continues:

> Because the stream and daylight are similar, the whole formed is homogeneous, and the motions caused by the stream coming into contact with an object or an object coming into contact with the stream penetrate right through the body and produce in the soul the sensations which we call sight. But when the kindred fire disappears at nightfall, the visual stream is cut off; for what it encounters is unlike itself and so it is changed and quenched, finding nothing with which it can coalesce in the surrounding air which contains no fire. It ceases therefore to see and induces sleep.[xx]

Typically, Aristotle disagreed. It made more sense to him to think of just one source of light, the object seen. Light from the object impinges on the eye, causing the sensation we call "seeing the object". This is an "intromission" theory of vision because the light goes *into* the eye to produce vision. Aristotle held to this view fairly consistently with one lapse. In analyzing the rainbow, he slipped into the extramission view, probably because his treatment at this point was rather mathematical and rays coming out of the single point of the eye are much more manageable than rays coming from just anywhere on the object.

Light itself, Aristotle argued, is not an entity or material substance but a state of the medium between an object and its viewer. The observed object, he thought, changes the medium, the change of which is transmitted to the eye to produce vision.

In the next century, the great geometer Euclid turned his gaze on light and adopted an extramission theory. In his *Optics*, he subtly altered Plato's understanding of the emitted ray. Calling attention to the fact that one often fails to see an object in plain view, Euclid argued that the emitted ray must be a beam and not a diffuse emanation, as Plato appeared to think. Many individual rays were emitted from the eye in a visual cone to form the beam. In Euclid's view, only when the visual cone struck the object directly was the object perceived. Of course, Euclid was quite pleased to work with visual rays that travel in straight lines and could be treated readily with the theorems and proofs of his geometry. The *Optics* became a very influential book in the history of thinking about light and vision.

Not surprisingly, the atomists had a quite different idea of vision. They were clever enough to invent an entirely original form of intromission theory. To these complete mechanists, the *active* seer was not to be countenanced. Seeing was a mechanical process in which the objects seen produce vision in the seer by husks or thin films in the shape of the object projected to the eye by the object. In the atomist view, people are not active but objects are! As atomists like Lucretius explained[xxi] these husks (*simulacra* or *eidola*) were thought to peel off the object and fly to the eye of the beholder. Light for the atomists was, like everything else, a matter of mechanics, of bodies in motion.

Their version of vision, incidently, prevented the atomists from being complete atheists. The fact that people have thoughts about the gods requires, in this view, the real existence of the gods. Without real gods to project *simulacra*, no one could think of the gods! The gods of the atomistic world were placed in an inaccessibly remote part of the universe, projecting simulacra at us from that vast distance.

How the *simulacra* of large objects such as trees or distant mountains could enter the eye posed a nasty problem for the atomists. Their opponents took such problems as proof of the absurdity of the view and atomistic explanations of vision were never popular. The atomists defended their view by pointing to the tiny images of the world that can be seen reflected from the dark pupil. Surely these images are the *simulacra* of the world!

A confusing aspect of the atomist view is that they also thought of sunlight as composed of minute atoms which, having to come to us through the resisting air, traveled more slowly than other atoms which moved in unresisting lock-step motion "down" through the universe. Apparently we see the Sun by the light it emits but other objects by the films they emit. That atoms of light move more slowly than other atoms seems not to have puzzled the atomists. Nonetheless, theirs is the first suggestion that light has a finite speed, even though it was a speed less than that of material objects.

The atomists also had an explanation of magnetic *attraction*. They believed streams of tiny particles, effluvia, flowed out of the lodestone toward the iron, forcing the atoms of air out of the way and creating a void. The air atoms on the other side of the iron struck the iron and propelled it into the void and toward the lodestone. As to why only iron was thus moved, Lucretius gave a wholly specious argument based on a supposed intermediate nature of iron between wood, through which the effluvia easily passed, and gold, which was too dense to be moved. Magnet *repulsion* was unexplained.

The astronomer Claudius Ptolemy also wrote a book on light and vision called **Optics**. No complete version of it is available but fragments of it were familiar to medieval scholars such as Roger Bacon and Witelo (c.1232-c.1275, first name unknown). Like Euclid (and probably for the same reasons), Ptolemy favored an extramission theory. The most notable feature of this work is the heavy use of experiment and observation. Ptolemy used specially made instruments to elucidate the behavior of reflected and refracted beams of light and devised experiments to test his theories. He showed experimentally that the angle of reflection equals the angle of incidence and was able to make approximately correct quantitative statements about refraction. A fully correct law of refraction, the bending of light rays on entering a different medium, eluded him but it must be noted that he was unfamiliar with the sine function on which the correct law depends. (Throughout his works, he used the now obsolete chord function in place of the lacking sine function.)

Ptolemy was not the first to state the law of reflection. Much earlier, Hero of Alexandria not only knew the law but also once again revealed his prescient physical intuition by noting that the law resulted if one assumed light always takes the shortest path from source to observer.

The last of the classical Greek scholars to contribute to theories of vision was the physician Galen. His contribution was basically negative because it was not only erroneous but, due to his prestige, it became an obstacle to forming a better view. Galen did not understand the function of the retina and mistakenly concluded that the visual image is formed in the "crystalline humor" or lens of the eye. He also spun out a wildly speculative account of the origin of Plato's fiery, visual stream. Ingested food, Galen taught, was transformed by the liver into "natural virtue" which in turn was transformed by the heart into "spiritual virtue". The brain changed this latter substance into a luminous wind of which a part became the visual stream.

At the end of the classical period there was no unified, monolithic theory of vision and light to pass on to succeeding generations. That was probably fortunate since so much of what was there was wrong. Later scientists had to sort out for themselves what light itself might be, how vision works, and whether or not the eye emits light. The legacy of classical Greece was not entirely negative. The mathematical analyses of Euclid and Ptolemy and Ptolemy's experiment-oriented approach would both be put to good use later on. The intromission theories had the potential to lead to distinguishing light from vision and the work on refraction would eventually make possible a better understanding of the action of the eye in producing images.

The Interlude

Alhazen made a genuine advance in theories of vision. Suspicious of the classical extramission theories he had inherited, Alhazen marshaled good reasons for opposing them. If light flows out of the eye, he said, how is it that looking at bright objects like the Sun can hurt the eye? Only if light flows from the Sun to the eye does the injury make sense. Even if the eye is not hurt, it is affected. If one looks for some time at a fairly bright object and then closes one's eyes, a *dark* image of the object is seen against a contrastingly light field. This suggests that the light of the bright object impinges on the eye to generate an image of the object. Furthermore, he asked, how could a visual ray from the eye almost instantly fill the vast heavens with light as one looks on them?

Alhazen, however, had philosophical commitments that constrained how he could respond to these perceived problems of the extramission theories. We are told that in his early years he had intently studied Islamic theology and had been dissatisfied with the disagreements he encountered. He resolved, therefore, to spend his time on more resolvable issues such as those he found in mathematics and, to a lesser extent, in science. But he came to his science with the conciliatory view that all truth is one, that theological truth and secular truth, rightly understood, cannot contradict each other. Thus, his inclination was to try to reconcile differences between the extramission and intromission theories. His book, also called *Optics*, therefore applied the precise, geometric techniques of the extramission theories of Euclid and Ptolemy to create just as precise an intromission theory of vision.

His most important decision toward this end he may have learned from the earlier Arab writer Abu Yusuf Ya Qub Ibn Ishaq al-Sabbah (c.801-c.866). Al-Kindi, as he was known in the West, apparently realized the power of the extramission theories lay in the mathematical precise of the visual cone emitted from the eye. The well-defined cone led to definite, clear answers about the possible behavior of the cone. In intromission theories, on the other hand, a light cone emanated from every point on the object seen, so intromission theories had to cope with an enormous tangle of rays arriving at the eye from the object. To simplify the situation, al-Kindi looked for a single, unique ray from each point of the object to the eye. The only possible choice (and thus, he thought, the *right* choice) is the ray from the point on the object to the center of the eye which traverses a radius of the eye. With this simplification, Alhazen succeeded both at producing a precise intromission theory and in finding a compromise between the two ancient views for which he had so much respect.

Building on the work of Ptolemy in reflection and refraction, he devised experimental verification of the law of reflection and explained refraction ingeniously and influentially with the assumption that light travels more slowly in dense media. The assumption is correct though he had no means of verifying it. He also improved on the experimental aspects of Ptolemy's work, using sighting tubes, strings, dark chambers, and instruments calibrated for measuring angles of refracted rays of light.

Alhazen also gave the first clear account of the *camera obscura* (pinhole camera). The name means "darkened chamber", a reference to the most obvious feature of the device. An enclosed box with a small hole in one wall and a translucent screen on the wall opposite the hole, the *camera obscura* produces sharp images of the bright outside world that can easily be seen by viewing the translucent screen from behind. Most importantly, the images are always up side down.

The images, even the fact they are up side down, can be readily explained assuming light travels in a straight line to the hole and beyond it into the chamber. The images are sharper but dimmer for small holes. This too makes sense because the small hole, like al-Kindi's model of the eye in vision, admits only a single ray from each point on the object. Notably, atomic theories of vision are not suited to the pinhole camera. How could the *simulacrum* from a large object get through the pinhole? Again, how can the *simulacra* from two objects pass through the hole simultaneously without tangling and giving a blurred image?

When schools and scholarship began to flourish in the West, both the extramission and intromission traditions came to light and were taught. Translators like Gerard of Cremona and James of Venice (fl. 1136-48) made most of the writings of Aristotle, Ptolemy and Galen available in Latin as well as the works of Al-Kindi, Avicenna and Averroes. Men like William of Conches (d. after 1154) strongly advanced Platonic thought, including the extramission theory of vision. William apparently taught both at Chartres and Paris before becoming tutor to the future Henry II of England and is a good example of how widely even one man could spread the new teachings.

Thus, when Robert Grosseteste began to study optics (or *perspectiva* as it was called in medieval Latin), a great many sources were available to him. The *Timaeus*, the *Optics* of both Euclid and Ptolemy and the writings of Aristotle and al-Kindi were all familiar to Grosseteste. The one serious lack in sources was Alhazen's *Optics* which had not yet been translated. Nevertheless, Grosseteste produced one of the world's most influential books on light and vision. *De Luce* (*On Light*) expresses Grosseteste's profound conviction that light is the basis both of the material and of the spiritual world. Greatly impressed by al-Kindi's idea that everything radiates light, Grosseteste declared that all matter forms from condensations and rarefactions of light. But, for him, the order of the material world was based on numbers and geometry.

The experimental methods of Ptolemy also impressed Grosseteste, encouraging him to become one of the early advocates of experimental and inductive ways of gaining knowledge of the material world.

Grosseteste's student, Roger Bacon, had the advantage over his teacher of having the *Optics* of Alhazen available to him. Bacon accepted most of the ideas of Alhazen and became a major conduit for the Islamic scientific tradition into the West. Like Alhazen, Bacon was inclined to reconcile the ideas handed down to him. Favoring the intromission theories of Aristotle and Alhazen, Bacon made room for extramission ideas by giving the visual stream of Plato, Euclid and Ptolemy the task of preparing the medium and the eye for vision.

The works of Alhazen and Ptolemy strengthened Bacon's tendency, inherited from Grosseteste, to expect experiment and observation to supplement knowledge gained from revelation and deduction. This experimental preference also appeared in the contemporary work of Witelo and in the most popular optics text of the Middle Ages, John Pecham's (d. 1292) *Perspectiva Communis*.

Another continental contemporary of Grosseteste and Bacon also promoted experiments, this time in the study of magnetism. Pierre Pelerin de Maricourt published an account of his magnetic experiments under the name Petrus Peregrinus in 1269. His most influential experiments were carried out with a terrella ("earthkin" or "little earth"), a piece of lodestone shaped into a sphere. Moving a small iron bar around on the surface of the terrella, the bar directions traced out lines of longitude. The intersections of the lines he called "poles" by analogy with the poles of the lines of longitude of the earth. Thus arose the term "magnetic pole".

The Renaissance

The new attitudes that arose with the Renaissance are perhaps nowhere more dramatically illustrated than in the study of light and vision because the Renaissance was in many ways an artistic phenomenon. The major figures we associate with that period are almost exclusively artists. In the works of Leonardo da Vinci, Albrecht Durer and Filippo Brunelleschi, a new way of doing art and of thinking about art is apparent. Part of the new art was a new vision of light; light the revealer of external phenomena as opposed to the light of internal illumination. A demand for seeing and portraying nature as it "really" is, for art as an imitation of nature, arose. So much effort was expended on the new way of doing art that very little lasting science was done at this time. The science came after the enthusiastic burst of artistic activity. But many suggestive ideas were considered at this time. For example, seeing the eye as a chamber, Leonardo da Vinci made the bold suggestion that the eye is a pinhole camera. Though it seemed impossible because the images of the eye are surely not up side down, Leonardo's suggestion had a fascination of its own that made it influential beyond reasonable expectation. We will soon have the opportunity to see Kepler and Descartes make impressive use of it.

The New Science

In 1589, Giambattista della Porta (1535-1615) showed that the new tool of navigation, the compass, was unaffected by onions and garlic. Using careful scientific procedures such as breathing and belching on the compass after having eaten onions and garlic, he found no change in the compass. He also rubbed the compass needle with extracts of these potent vegetables to no effect. To us, these procedures are almost too simple to deserve being called experiments. However, at the time, Porta's activities were uncommon. Rumors that onions and garlic were deleterious to the function of the compass were current (though Porta tells us that the experienced sailors he queried on the subject were scornful of the idea). Yet, no one before Porta thought to check them out.

Porta also checked on the magnetic theories of Diogenes of Apollonia (c. 460 BC). Diogenes believed that the "dryness" of the magnet seeks to feed on the "moistness" of iron, causing the familiar attraction of the magnet for iron. Porta reported in his *Natural Magic* that a lodestone he buried for many months in iron filings did not gain significantly in weight. He judged Diogenes' idea doubtful.

As in other areas of science, developments in optics occurred rapidly after 1600, largely driven by new and more precise information gleaned by more careful experiments and observations. William Gilbert's *De Magnete* opened the century impressively. Following Petrus Peregrinus, Gilbert too constructed a spherical terrella of lodestone for his studies and verified the existence of the magnetic poles. Not entirely free of the animism of other centuries, Gilbert argued his terrella, having the most perfect form, the shape of the earth, was therefore a true "offspring" and model of the earth. The importance of this attitude is that it made him see the Earth as simply a large version of his lodestone terrella. When he broke the terrella and found that magnetic attraction made the pieces cohere tightly just as if the sphere had never been damaged, he then mistakenly concluded the Earth too was held together magnetically. This led him to argue that rotation of the Earth would not cause it to fly apart, a helpful but wrong idea.

Moving a small compass over the surface of the terrella, Gilbert found readings that replicated the compass behavior seen by mariners. Again, the Earth and the terrella were alike, magnets. Here, then was the explanation of the behavior of the compass.

Gilbert noted that iron can be magnetized and will then attract iron just as the lodestone does and he called attention to the fact that magnetism, both in iron and in the lodestone, can be destroyed by heat. He also explicitly called attention to the repulsion of like magnet poles and the attraction of unlike poles.

Unaware of electrical repulsion, he used the existence of magnetic repulsion to distinguish magnetic forces from electrical forces. Like the atomists, he attributed electrical attraction to effluvia but, seeing no evidence of deflection of magnetic effluvia, he denied the truth of any effluvia theory of magnetism, explaining it instead in terms of rays of magnetic virtue emanating radially from the center of the magnet. He explicitly and erroneously denied Porta's finding that the magnetic rays emanate from the poles.

Gilbert also added to knowledge of electrical action, giving a long list of materials beside amber that attract chaff when rubbed. The list included many gems such as diamond, jet, rock crystal, sapphire, opal, amethyst, and beryl as well as glass, paste gems, sulfur, and sealing wax!

As we have noted earlier, Kepler was much impressed with Gilbert's work and was inspired by it to suggest a magnetic force from the Sun held the planets in their orbits. Kepler also dilated on Leonardo's idea of the eye as a *camera obscura*. He analyzed the *camera obscura* and showed with careful geometric arguments how it formed its inverted image. The eye he saw as a similar object with the pupil as the pinhole and the retina as the screen. Here he broke with earlier writers who thought the image was formed at the center of the eye. Both the analogy with the *camera obscura* and Kepler's physical sense were behind the break. He argued that the image had to be formed on the first opaque object the light encountered and it could not be formed in the transparent center of the eye.

Lacking a precise knowledge of the law of refraction, Kepler was unable to attain a detailed understanding of how the cornea and lens of the eye, the refractive elements, assist in forming images when the dilated pupil is too large to be a true pinhole. However, for Kepler and his contemporaries, the most disturbing aspect of his retinal theory was the forced conclusion that the image on the retina is inverted. Kepler himself, refusing to allow the problem to demolish his theory, simply said he, as an "optician" left the problem of how the mind made use of the inverted image to the "philosophers".

Once again, Kepler acted as a watershed. His distinction between the physics of how the image is formed and how the mind treats that image was the critical one for separating the behavior of light from the actions of the observer of the light. The theory of vision was finally separated from questions about the nature and behavior of light; light was now an entity in its own right and no epiphenomenon of the mind or mere characteristic of the medium. As with the Mertonians' distinction between describing and explaining motion, a major problem was now broken down into simpler, more tractable pieces. Progress would now be made more easily.

Galileo too turned to the study of optics, most notably in his production and use of the Galilean telescope on which his study of the heavens was based. His greatest contributions to optics came when he turned to questions of the nature of light. He advocated an atomic theory of light, using the dangerous word "atom" in this particular case.

He also attempted to measure the speed of the atoms of light. The measurement involved Galileo and an assistant, each with a shuttered lamp. They set up the lighted lamps facing each other on hill tops from which each observer could see the light of the other. At night, Galileo unshuttered his lamp and his assistant, seeing the light from Galileo's lamp go on, unshuttered his lamp. Galileo timed the interval between opening his shutter and seeing the light from his assistant's lamp. Unfortunately, the high speed of light made the time interval basically a measure of human reflex time and the experiment produced no useful value of the speed of light beyond showing the speed had to be very great.

Galileo also took note of the newly discovered phosphorescence of certain minerals (BaS in this case). Exposed to bright sunlight, these minerals then glow in the dark without giving off noticeable heat. The mineral was first called "solar sponge" although the name was soon recognized as misleading because the light emitted is of a different color than sunlight. The idea that light and heat (fire) are connected is very ancient. Galileo was fascinated with this light divorced from heat. The phenomenon encouraged him to distinguish heat from light and, though he held an essentially atomic theory of both, he thought it necessary to speak of atoms of light but "fiery minims" of heat. Thus we discover another situation where a clear distinction made room for scientific progress.

Another Italian, a Jesuit named Nicolo Cabeo (1585-1650), made an important but isolated observation around 1630. Cabeo noticed an effect that others had missed: two objects may initially attract each other electrically but, if they touch, they thereafter repel each other. Here was the electrical repulsion Gilbert thought did not exist. Cabeo was a strong anti-Copernican. He was motivated by the fact that Kepler and Galileo, both Copernicans, used Gilbert's ideas in their work. Cabeo evidently hoped to discredit Copernicanism by discrediting Gilbert! In this hope he was greatly mistaken. Gilbert's work was in no sense critical to the work of either one. This is a, by no means unique, case of scientific progress made from the worst and most foolish of motives.

About this time, Descartes, the deductive philosopher, actually carried out experiments in optics. Obtaining an ox's eye, he partially scraped the retina from the back of the eye. Then he viewed some brightly-lit objects through the eye. In agreement with Kepler's theory, he saw a sharp, ***inverted*** image of the objects on the partially scraped retina. So Kepler was right; the eye forms inverted images and the brain must be reinverting them!

Snell had discovered the sine law of refraction (which now bears his name) in 1621. Consequently, Descartes, who was actually the first to publish Snell's law (1638), was able to include the qualitative aspects of the refraction of the cornea and lens in his analysis of the retinal image. A quantitative analysis of the actions of these refractive surfaces was still too difficult. Even in his evaluation of the rainbow, Descartes had to model the raindrop with a prism in order to be able to be quantitative, but this model cannot produce the secondary rainbow, let alone its inverted color order. Thus, Descartes' calculations were still less precise than the earlier experimental results of Theodoric on the production of rainbows by water filled spheres.

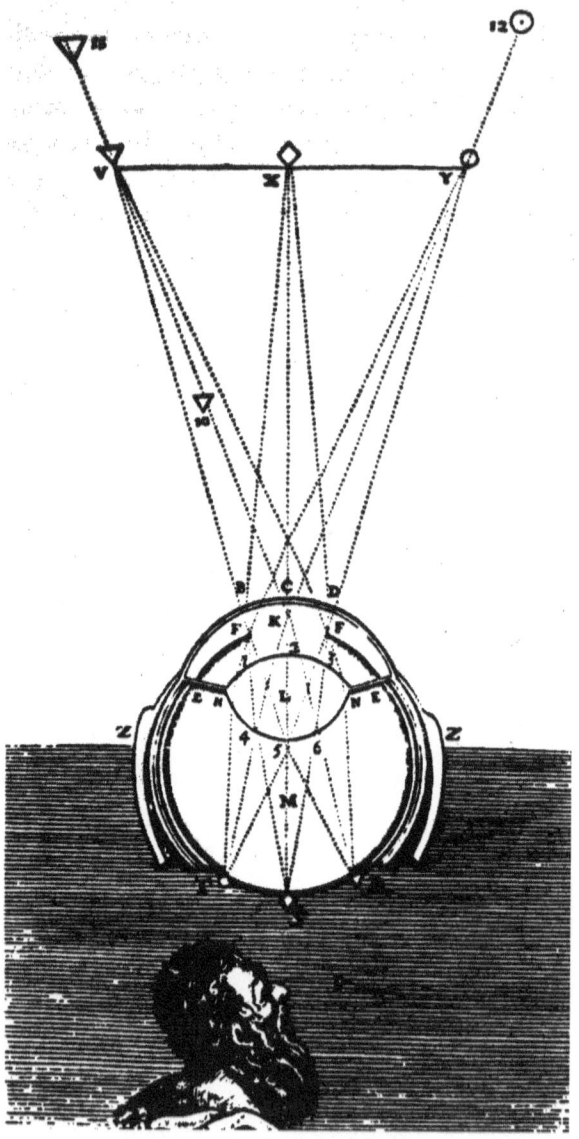

Figure 30. Descartes examines the inverted image on the retina of an ox-eye.

Descartes was able, however, to improve understanding of the rainbow and the splitting of light into colors by the prism. Snell's law can be written

$$n_1\sin(i) = n_2\sin(r)$$

where i and r are the respective angles of incidence and refraction at a surface. The n_1 is the refractive index of the first medium in which the incident light travels and n_2 is the refractive index of the medium into which the refracted light penetrates. Descartes realized that, since r is different for different colors (that is, the colors separate), n_2 must be different for different colors.

As to the nature of light, Descartes remained Aristotelian. Since, like Aristotle, he held that the universe is a *plenum*, he also held that light was a characteristic of the medium, the *plenum*. Resisting the idea of atoms of light in motion, Descartes regarded light as a tendency toward motion of the *plenum*, a tendency that traveled infinitely fast.

Impressed with Gilbert's work on magnets, Descartes was moved to tamper with it. Replacing Gilbert's rays of magnetic virtue with "threaded parts", particles of magnetic attraction which, discouraged by Gilbert, he could not call "effluvia". With this idea he claimed he could explain all known magnetic phenomena. His scheme was this. There are pores through the Earth in pretty much parallel lines which touch the surface at the magnetic poles (he understood magnet poles as sides of the Earth rather than as points in the Earth). There are two types of threaded parts; one only enters at the north pole and exits at the south and the other does the reverse. The threaded parts constantly travel through the Earth and around it in their ordained directions. Since travel through the air is unnatural and difficult they will preferentially travel through a lodestone or piece of iron if available since it also has magnetic pores. In fact, threaded parts can even abandon the Earth and circulate exclusively around and through iron or lodestone, thus magnetizing the host magnetic material. In essence, Descartes revived the effluvia theory of the atomists.

A contemporary of Descartes and a fellow Frenchman, Pierre de Fermat (1601-1665) was critical of Descartes' metaphysical deduction of Snell's law. Instead, taking a cue from Hero's principle that light follows a path of minimum *distance*, Fermat derived Snell's law from a new minimum principle. Now called Fermat's Principle, this idea is that light travels the path of minimum *time*. Hero's principle of minimum distance was obviously wrong for refracted rays, the bent path traveled by a refracted ray is longer than a straight path would be. Fermat's ideas also required that light must slow down on entering a medium with a greater value of refractive index whereas Descartes' ideas led to the opposite conclusion.

As to developments in electricity, in the 1660's von Guericke turned his considerable talents from pumps to electricity and succeeded in producing a machine that generated a continuous supply of electricity through continuous frictional rubbing. This was the first electric generator but, at first, it was basically a novelty and toy.

Another piece was added to the puzzle of electricity and magnetism in 1681. It was widely reported that a ship headed to Boston was struck by lightning. When the excitement died down, it was discovered that the compass had reversed direction. The north-pointing needle now pointed south. The navigator made the necessary mental adjustments and the ship arrived safely in Boston. But, it was now clear that electricity could also alter the magnetic state. There must be a connection between electricity and magnetism: but what could it be?

By the middle of the seventeenth century, progress was being made in understanding sound as a wave. Boyle's work with the vacuum demonstrated beyond doubt that while sound does not travel through a vacuum light does. Obviously, sound requires a medium; that implied that sound is a disturbance of the medium. In fact, it was becoming increasingly clear that sound is a longitudinal wave. Boyle noted that electrical forces also propagate through a vacuum but he never connected that to the similar behavior of light.

Now, while the view of light as a stream of particles or atoms appeals powerfully to the imagination and can explain much about light, it does not explain all the behavior of light. For example, Descartes rejected the particle theory because he thought the particle beams from two people looking directly at each other would necessarily interfere with each other, thereby distorting what each person saw. It occurred to several people that if sound is a disturbance of a medium, perhaps Aristotle's idea of light as an effect of an object on a medium can be made to work. As early as 1664, Robert Hooke proposed a wave theory of light. Fourteen years later, Huygens constructed a much more thorough wave theory although he did not publish it until 1690.

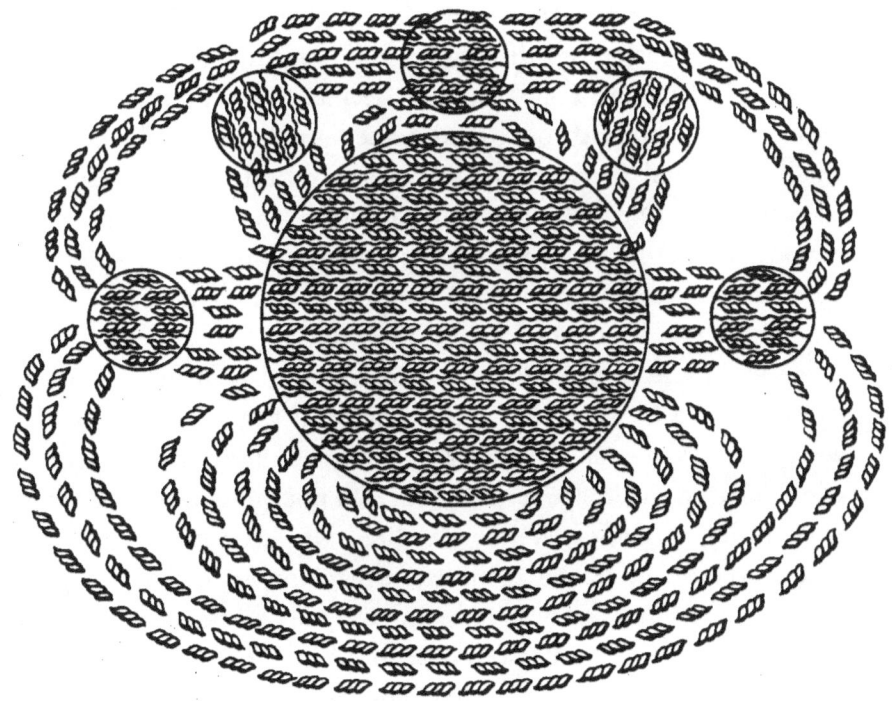

*Figure 31. Descartes idea of how magnetic **effluvia** create magnetic forces.*

The most important feature of Huygens' theory was that he produced a mathematical analysis of how a wave front propagates. Others were thus enabled to use the theory to predict wave behavior in different situations. A major accomplishment of the theory was its explanation of the birefringence of calcite crystals (called Icelandic spar at that time). Objects viewed through a clear calcite crystal appear double and the double images rotate around each other as the crystal is rotated. Unfortunately, his theory could not explain the differences between the two images (we now know the images have different polarizations). In 1717, Newton showed this failure was a consequence of assuming light is a longitudinal wave.

It has long been recognized that all waves bend around corners. We say they diffract. Since light was not known to diffract, Newton rejected a wave theory of light. When the Jesuit scientist Francesco Maria Grimaldi's (1618-1668) observation of diffraction of light around a rod in a beam of sunlight was published in 1665, Newton decided it was most likely due to a periodicity of the ether through which the light traveled. The effect was very small compared with known waves like sound and Newton's view seemed reasonable. Newton was also willing to consider the possibility that periodic effects like Newton's rings, were products of the details of refraction and reflection. Hence, he made his famous observation that light has "fits of easy Transmission and Reflexion".

No one was more prominent in the optical studies of the seventeenth century than Isaac Newton. His involvement in the subject began with the Newtonian telescope. All earlier telescopes suffered from chromatic aberration. That is, the image seen was fuzzy, a not quite overlapping set of images of different colors. Newton realized the aberration was due to Descartes' variation of refractive index with color. If refraction was the problem, then the solution lay in reducing the use of refraction in forming the images. Since reflection does not produce chromatic aberration, Newton designed and constructed a telescope with a focusing mirror as the main light gathering and magnifying component. He ground the mirror himself. So great was his skill that instrument makers initially failed to produce functioning copies when demand for the telescope arose.

This work was done in Cambridge where Newton held the Lucasian chair of mathematics. When word of the new telescope reached the Royal Society in London, they sent a request to examine it. Newton obliged, sending a twenty five inch long version which was received with high enthusiasm. On the strength of this feat, Newton was made a member of the Society and he soon communicated to them the results of his further work in optics. Eventually, this material was published as Newton's *Optics*.

A highlight of this work was Newton's demonstration that white light is composed of the colors of the rainbow. In his *experimentum crucis*, Newton first used a prism to spread white light into the rainbow then, with a second prism, he recombined the rainbow of colors into white light. He thus settled a long running debate on the colors of the rainbow. Either the colors were components of white light, as Newton showed, or they were products of the prism.

The colors, Newton imagined, were tied to the size of the light particles with the smallest particles creating the sensation of violet light and the largest making us see red. This ordering is a consequence of Newton's ascription of mass and inertia to particles of light. Since red light is bent the least by a prism, red particles must be the most massive, presenting the greatest resistance to the deflecting powers of the prism (Newton assumed light particles to have inertia).

These ideas did not go uncriticized. If light has mass, the Sun must be losing mass at a great rate. The answer was that the mass of light particles must be minuscule.

In 1676, Ole Christensen Romer (1644-1710) produced the first reasonable value of the speed of light with an ingenious calculation. He noticed variations in the times taken by the moons of Jupiter to undergo eclipse by the shadow of Jupiter. He had the inspired, creative thought that the variations were due to the motion of the Earth around the Sun which can lead to substantial changes in the distance the light must travel from the moon to an earthbound observer. Using the then available distances within the solar system, Romer got a value within 25% of the correct value (the speed of light is now a standard value that does not depend on measurements).

By the end of the seventeenth century, new information had pushed optical theory to more precise ideas of what light might be like and to new heights of mathematical sophistication. The old particle theory of light was dominant, in keeping with the mechanism of the times, though the *simulacra* of the old atomism had fallen by the wayside. The medium theory of Aristotle had been transformed into a wave theory. A means of deciding between the two types of theories was known; wave theories required light to slow down in optically dense media (high values of refractive index) and particle theories require just the opposite. However, techniques for accurately measuring what was increasingly understood to be the very high speed of light were not yet up to this critical task.

The Eighteenth Century

If the publication of Gilbert's *De Magnete* in 1600 opened the seventeenth century, then we can perhaps allow a lecture by John Keill to open the eighteenth century. Keill was Savilian professor of astronomy at Oxford and, in his eighth lecture of 1700, he announced that just as heat can destroy magnetization in lodestone or iron, so too can a blow, sharply delivered. He speculated that both situations alter the "internal structure" of the magnet. He further remarked that effluvia theories provide no help toward an understanding of such effects.

With the weight of Newton's authority and prestige squarely behind the particle theory of light, developments in that area were slow in coming. However, electricity finally became an area of active investigation in the eighteenth century.

Fully a century after the work of Cabeo, an obscure Englishman named Stephen Gray (d.1736) began to build on it, probably unwittingly. Benefiting from von Guericke's electric generator, he discovered that metals could convey electricity over considerable distances. Gray also noticed that electric charge resides only on the outside of bodies.

Four years later, in 1733, the superintendent of the French Royal Botanical Gardens, Charles Francois duFay (1698-1739), learned of Gray's results. This led him to distinguish materials into electric conductors and electrical insulators. Cabeo's observation of electric repulsion then encouraged duFay to advance the theory that there are two types of electricity, vitreous and resinous, which attract each other but repel their own kind. His successors, especially Jean-Antoine Nollet (1700-1770), turned to the view that, if electricity flows like a fluid, then it must be a fluid. Hence, the two-fluid theory of electricity arose.

A valuable technical advance became available around 1746, the Leyden jar. A glass jar coated inside and outside with metal foil, with a cork in the mouth pierced by a metal rod to conduct electricity to the inner metal coating, the Leyden jar collected and stored electricity from rubbed objects. One of the claimants to the invention, Pieter van Musschenbroek (1692-1761), was a resident of Leyden, hence, the name.

Another useful device invented in the middle of the century was the electroscope. Gray had used an ivory ball and feather on long threads to detect electricity. When electrically charged, the two objects repel each other and moved apart. In the electroscope, the ball and feather were replaced with two very delicate metal foils. The degree to which they separated gave a crude estimate of the degree of electrification.

About this time, the American scientist Ebenezer Kinnersley (1712-?) became an advocate of the two fluid theory. His fellow Philadelphian and co-experimenter, Benjamin Franklin (1706-1790), had a different opinion. Franklin saw no need of two fluids when a single fluid in excess or deficit would do as well. Unfortunately, Franklin labeled negative (the deficit) the type of charge that flows in most electric current. It was an arbitrary choice at the time but it continues to bedevil introductory physics students with a positive current going one way due to negative electrons going the opposite way.

Franklin is also justly famous for his kite experiment showing that lightning is electrical. The experiment was much more carefully done than is usually realized. Franklin did not fly the kite into an electrical storm, but when one was nearby he stood under a shed, and kept a length of dry line between himself and the kite. He stored electricity from the kite in a Leyden jar. Proving the experiment was risky, one Georg Wilhelm Richmann of St. Petersburg in Russia was struck and killed by lightning trying to repeat Franklin's result. Some of the major scientific journals of the day reported his autopsy results because so little was known of the effects of electrocution on the organs of the human body.

The kite experiment was only a small part of Franklin's program of studying electricity. He invented and used a grounded metal cage for protection from lightning and it is well known that he was an early advocate of lightning rods. He installed rods of his own design on several buildings and even, much to his wife's disgust, had a very special one in his house which had a small gap in it. Whenever lightning struck that rod, a powerful spark appeared across the gap.

Franklin's work on lightning was the prime impetus behind recognition of lightning as an electrical phenomenon; he removed it from the realm of things unpredictable and uncontrollable into that of things understandable and manageable. He speculated that St. Elmo's fire was also an electric discharge from the rigging and masts of sailing ships, reducing this phenomenon, too, to less awe-inspiring proportions.

Experimenting with and reflecting on the fact that electric charge resides on the outside of a hollow body, Franklin also made the useful and correct suggestion that the electric force inside such a body must be zero. His friend and correspondent, Joseph Priestley confirmed the idea experimentally in 1767 and made the further useful and correct suggestion that the electric force law might then be an inverse square law since gravity, obeying such a law, also gives zero force inside a hollow shell.

Franklin's one fluid theory had the advantage of emphasizing the conservation of charge, a major contribution on Franklin's part to the laws of electricity. On the other hand, the two fluid model led to fruitful connections between electricity and magnetism and to mathematical laws for the magnetic and electrical forces. By about 1780, Cavendish had used his gravitational torsion balance to verify that electric repulsions and electric attractions both obey an inverse square law with distance. However, he failed to report this and it was through similar work done in 1785 by Charles Augustin Coulomb (1738-1806) that the world learned of the inverse square law of electric forces. While he was at it, Coulomb confirmed earlier work showing that magnetic forces also follow an inverse square law in distance between poles. Thus, another similarity between electricity and magnetism (and gravitation) appeared.

With the Leyden jar available to capture and transport electricity, all manner of silly electrical parlor tricks came into vogue. Even such serious students of electricity as Franklin could not resist. Making people jump in surprise at the sensation of an electric shock led to the realization that electricity can cause muscle action. This led to further silliness with electricity viewed as a nerve tonic and stimulant. However, these developments had a scientifically useful side; Alessandro G. A. A. Volta (1745-1827) came to recognize muscles as detectors of electricity.

He was able to use this understanding when Luigi Galvani (1737-1798) reported a curious effect in frog legs. In 1786, Galvani had prepared some frog legs (complete with spinal column and crural nerves) for his studies in physiology and hung them on brass hooks. When he set the hooks on an iron support, the frog legs twitched. Galvani, mentally focused on the muscles, thought the muscles originated the action. Volta, predisposed to think of muscles as detectors of electricity, immediately recognized that the contact of the two different metals was somehow generating electricity, exciting the muscles. The metals were somehow producing an "electromotive force" (emf) to make the frog legs twitch. Today, in honor of Volta, we simply refer to the electromotive force as a *voltage* and we speak of "galvanic action" in corrosion phenomena.

By 1792, Volta had learned to make a "Voltaic pile" of many such junctions of two metals so as to greatly increase the effect over that of a single junction. Before the turn of the century, he had also discovered the effect could be made continuous if the junctions were made with, or at least moistened with, some acid. He had made the first electric battery. More importantly, Volta had begun the era of continuous electric current. Until that point, the only current flow available had been sparks of electricity that typically lasted only small fractions of a second. Unfortunately, the continuous currents were not yet constant. The experimentalist, working on a new phenomenon, wants to begin with the simple cases, with as many variables as possible held constant. Hence, the early Voltaic piles just were not good experimental devices. Detailed, quantitative study of current electricity had to await further technological advances.

The only notable results on light during the eighteenth century came from Euler. The greatest mathematician of all time in terms of volume of work and, perhaps also of quality of work, Euler was one of the very few scientists of the century who could afford not to be intimidated by Newton's legacy. He published his ideas in 1746 in his **New Theory of Light and Color** and came down squarely in favor of light as a vibration. Nonetheless, even with his enormous mathematical powers, he could not give a satisfactory explanation of diffraction.

By the end of the eighteenth century science had become fairly sophisticated. This trend has continues in the science of the nineteenth and twentieth centuries. Increasingly, scientific equipment became more elaborate and expensive. The solitary worker gave way to large laboratories with a division of labor. Concepts become subtler as the phenomena become known in greater and greater detail. Consequently, our story becomes noticeably more difficult as we move into the next centuries.

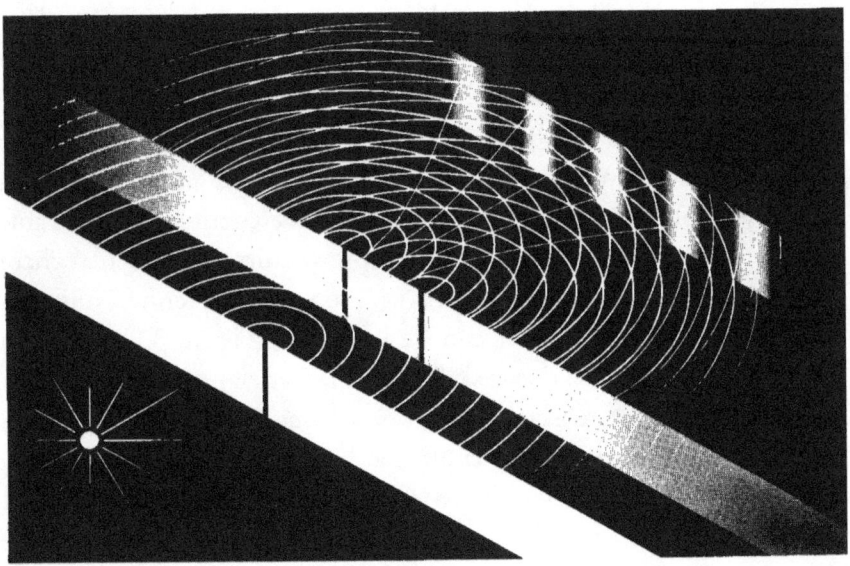

Figure 32. Young's double slit experiment. Light from the source at the left goes through a single then a double slit, creating an interference pattern on the final screen on the right.

The Nineteenth Century

Following Euler's ideas, Thomas Young argued that Newton's rings could only be explained in terms of interference of waves. After all, Newton's remark of "fits of easy Transmission and Reflexion" was made with reference to the rings. Seeking a way to show definitively that light is a wave, Young performed his justly famous double slit experiment (Figure 32). His idea was that the two slits acted as separate sources of light waves which, beyond the slits, combined their light as waves, giving areas of bright light where none should have appeared if light is made of particles which only travel in straight lines. His idea was that at places where the peaks of the two light waves coincided, the screen would be bright. Where one wave was at a peak and the other at a trough, the two disturbances should nullify each other and the screen would be dark. The results were exactly as expected. Young even checked the result of blocking one slit; the pattern disappeared as expected.

Alas! the results were not as definitive as Young had hoped. The defenders of a corpuscular nature of light put their heads together and were able to invent an explanation of Young's results from within a corpuscular theory.

For most scientists, the break came in 1808 when Etienne Louis Malus (1775-1812) showed that light can be polarized (Figure 33). Using birefringent calcite crystals, Malus, a skilled mathematician, extended Huygen's work on birefringence to show that Newton's criticism of Huygen's wave theory was wrong (Newton had explained birefringence by assuming light particles had flat sides!). Then, in 1816, Fresnel and Arago collaborated to show that polarization could only occur with transverse waves.

If the demonstration of polarization did not convince nineteenth century scientists that light was a wave, one more convincing detailed was added to the picture. Fresnel had first ventured into science (he was a civil engineer) by submitting a paper in a contest held by the Paris Academy of Science. Unaware of Young's results and isolated in the provinces, Fresnel wrote a brilliant paper into which he wove the results of his own experiments on diffraction (the local blacksmith made instruments to his design) and a very original and sophisticated mathematical analysis. The judges were staunch Newtonians and, though they awarded the prize to Fresnel, some sought to discredit the wave theory on which the paper depended. Simeon-Denis Poisson (1781-1840), the eminent mathematician, was one of the judges. Extending some of Fresnel's results, Poisson showed that Fresnel's theory required that a bright spot of light must appear in the center of the shadow of circular object. The result seemed absurd and cast doubt on all wave theories of light. Arago, strongly supportive of Fresnel, did the experiment and showed that, indeed, the bright spot appears! The confirmed prediction of a totally unexpected phenomenon is the strongest supporting evidence a scientific theory can get. The tide turned in favor of light as a transverse wave.

Newton's corpuscles were forgotten but Aristotle's aether, revived by Newton, did not suffer the same fate. It was even more important now because, obviously, a wave needs a medium. What is a wave but a disturbance of a medium?

It is worth comment that it was primarily French scientists who worked to overthrow the particle theory of light. While most French scientists were Newtonians, nationalistic attitudes prevented French enthusiasm for English theories from being as strong as English enthusiasm. Thus, it was easier for French scientists to break away from the grasp of Newtonian bias than for the English. The reason it was French and not, say, Italian scientists is because of the Enlightenment tradition in France (or the Cartesian tradition, if there is a difference). These traditions had generated in France a scientific atmosphere and educational program of, at the time, unparalleled breadth. Because of it, England lost the place of scientific priority it enjoyed in the seventeenth century to France in the eighteenth century.

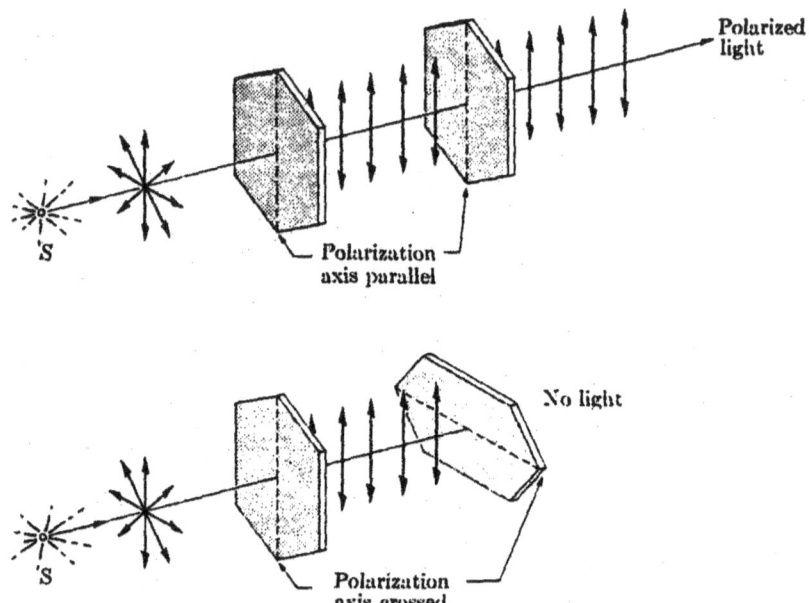

The light polarized by one tourmaline crystal passes through the second crystal in one position (upper diagram) but is cut off when the second crystal is turned through 90° (lower diagram).

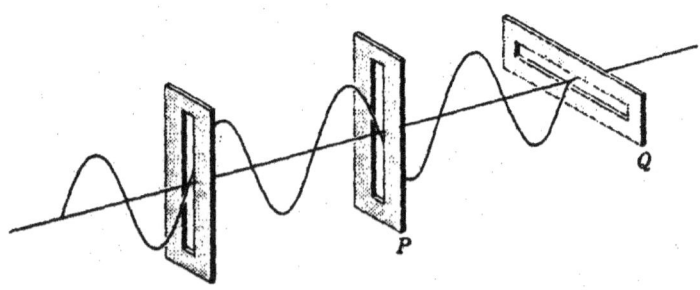

A slot lets a transverse wave in a string pass if it is aligned in one direction (*P*), but cuts it off when rotated through 90° (*Q*).

Figure 33. Only transverse waves can be polarized so light must be a transverse wave.

With the successes of Arago, Fresnel, and Young, the image of light as a particle was broken. In its place was erected a new image, light-the wave. Not until early in the next century was the particle image resurrected.

Rapid though not quite as revolutionary developments also occurred in the fields of electricity and magnetism in the early nineteenth century. The supply of electric currents of many seconds duration drawn from the Voltaic battery made possible the pursuit of the connection between electricity and magnetism. In Denmark, Hans Christian Oersted (1777-1851) began an experimental search for the connection in 1807 but was unsuccessful for a dozen years.

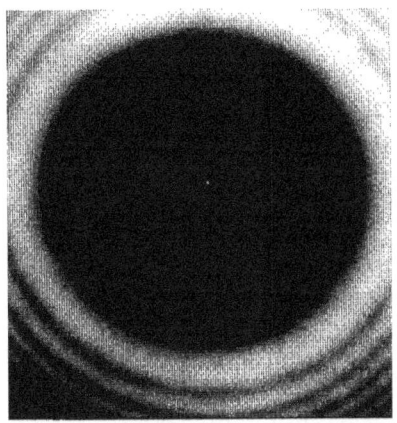

Figure 34. The shadow of a penny. Note the central bright spot.

In April of 1820, his efforts paid off. He had been working under the impression that the strongest magnetic effects should appear under static conditions with no current flowing to drain the electric powers. He was astonished when toying with equipment at the end of a lecture, his assistant got a compass needle to deflect by approaching it with a current bearing wire. More careful examination showed the compass needle turned so as to point at right angles to the wire! The current was producing a magnetic field at right angles to itself, not coming out of the wire but circling around the it.

Oersted immediately sent a memoir in Latin to scientific colleagues and societies around the world, triggering a decade-long orgy of scientific work and discoveries in the new area of electric and magnetism. Arago reported Oersted's result to the French Academy on September 11, 1820 and then quickly showed that electric current both attracts iron filings and magnetizes iron needles. Electric current acts like a magnet!

Exactly a week after Arago's report, Andre Marie Ampere (1775-1836) suggested to the academy that magnetism might be due to internal electric currents in the iron. This thought led on to the possibility that steel needles magnetized by the strong magnetism of many coils of current carrying wire should be more strongly magnetized than similar needles magnetized using a single wire. Ampere persuaded Arago to help him and they were quickly able to confirm the idea experimentally. Fresnel then wrote Ampere with the thought that the internal currents must be molecular rather than macroscopic. He was on the right track but, with no clear picture of the structure of molecules and no hint of the laws governing their behavior, Fresnel's idea lapsed for more than a century.

By a series of ingenious experiments, Ampere, Jean-Baptiste Biot (1774-1862), and Felix Savart (1791-1841) worked out the details of forces between current-carrying wires and both magnets and other current carrying wires. They were able to express their results in the laws named for them, the Biot-Savart law and Ampere's law. Because currents cause magnets to turn at right angles, these two laws have to be expressed in terms of the calculus of vector quantities and are quite difficult to handle. In fact, this right angle business is a major significance, though that was not immediately obvious. Newtonian physics was framed around forces that came radial out from a source. Forces coming out of their source at right angles, forces encircling their source do not fit Newtonian physics. Magnetism is inherently non-Newtonian.

Technologically, by 1825, currents through wire coils were being used to produce strong magnetic fields. The electromagnet was born.

On the purely mathematical side, several investigators, but especially Poisson and Laplace, contributed to the development of an elegant and powerful mathematical vision which convincingly explained much of what was known of electrostatic and magnetostatic situations. These terms refer to situations where electric charges and magnetic sources are unmoving and unchanging. This work was very much an extension of the work of Lagrange and Laplace on gravitational and mechanical problems in which mass acts as a source of gravitational force. Poisson replaced mass with "potential", a quantity that was precisely defined but lacked clear physical interpretation until it was eventually connected with potential energy and voltage.

Unlike Ampere, who willingly made conceptual leaps to explanations at a deeper level, Poisson would not push potential past its mathematical definition to a more intuitive conception. Ampere was quite willing to suppose that all magnetism was the result of currents, even if that meant some of the currents could not be verified. It was the most parsimonious idea and it might lead to other physical intuitions. Although Poisson had done his work assuming magnetic effects were due to a fluid, he would not speculate on the details of the nature of the fluid. Of the two, it is Ampere who was most like a modern physicist.

The 1820's also saw advances in the study of electric current. Thomas Johann Seebeck (1780-1831) discovered "thermoelectricity" in 1821. Looking for further connections between electricity and magnetism, Seebeck made a circle of metal. One semicircular side was copper and the other was bismuth. Such a "sandwich" of a wire of one type of metal spliced into a wire of another type of metal is called a thermocouple. When either junction was heated, a current flowed around the ring. Seebeck went on to show that all combinations of metal produce this "thermomagnetism", as he called it, with bismuth and tellurium giving the strongest effect. Here, then, was an alternate source of electromotive force (emf). Since the emf it produced was directly proportional to the temperature difference across the two junctions, the emf, of the Seebeck effect is extremely easy to control experimentally (i.e., temperatures are easy to control). A "thermopile" of many of Seebeck's thermocouples in series made the first highly reliable voltage source. The availability of thermopiles enabled Georg Simon Ohm (1787-1854) to carry out his groundbreaking investigations (1825-1827) of conduction of electric current in wires. That work led to Ohm's law of electrical conduction

$$V = IR$$

the fundamental equation of modern electrical studies. V is a voltage that drives a current, I, through a resistance R.

In 1849, Kirchhoff identified Poisson's electric potential with electromotive force. Surprisingly, to convert one to the other required multiplying one by the speed of light! What did light have to do with voltage?

With the idea of electric current producing magnetism well established, the reverse possibility, electric current produced by magnetism, became a serious area of investigation. By 1829, an American, Joseph Henry (1799-1878), had noticed that making or breaking the current in an electromagnet produces instantaneous but enormous voltage surges in the electromagnet. This is a case of what is now called electromagnetic self-induction. He reported this in 1832, too late. Michael Faraday (1791-1867) had already published the results of his experiments that showed mutual-induction as well. Electromagnetic induction, Faraday found, occurs in a loop of wire when the current it carries changes (self-induction) or when change or movement occurs in an external magnetic source (mutual-induction). The results are summarized in Faraday's law of induction. From this work springs the dynamo (Faraday made the first one) and electric power generation as we know it today.

Faraday became the dominant experimental physicist of the century, and his background had a profound effect on his work. He grew up in the slums of London, the son of a devout and serious-minded blacksmith.

Not strong enough to take over his father's business he was apprenticed, at age thirteen, to a bookbinder and used this contact with books to educate himself. Most importantly, he received no significant training in mathematics. Fortuitously, he came to the attention of Sir Humphrey Davy, the great chemist, who, despite the title, was also of modest background. It was Davy who trained him in experimental science. Because of his lack of training in mathematics, Faraday developed a powerful, visual, pictorial way of thinking about physical situations. The pictures and images he invented were so useful that they have come to be part of the standard mental equipment issued to budding physicists.

Faraday thoughts of a magnet as producing a magnetic field (a region where a magnetic force appears on compasses or other appropriate detectors). The magnetic field was composed of lines of force that can be seen by looking at the alignment of iron filings in the field. The lines curve from one pole to another and are more concentrated at the poles. In magnetic induction, lines of force are cut as a coil moves or as the lines themselves move in response to changing currents. The number of lines cut per second was then directly proportional to the voltage induced in the coil. The voltage then produced the induced current. Clearly, lines of force and fields may also be applied to gravitational and electrical forces.

Faraday, in fact, went on to advocate a non-material view of atoms as dimensionless centers of force. For him, space was filled with ether; atoms were points without size from which fields emanated. If the universe contained matter, it was the ether, not atoms! Thus was born the idea of force fields that has since played so large a part in our modern vision of the physical world.

The first result of the field idea was the expectation it created, in Faraday's mind at least, of a connection between magnetic fields and light. It took him years but, by 1845, he found the Faraday effect, where a beam of light polarized in a particular plane is found to have had that plane rotated on passing through a medium parallel to the direction of magnetization of the medium. A number of other such effects are now known.

Also in 1845, Faraday discovered that *all* materials can be magnetized, albeit weakly. Any material placed in a magnetic field acquires a very weak magnetization that is in the direction opposing the direction of the magnetizing field. This type of magnetization is called diamagnetism; it (and not, as Fresnel thought, ferromagnetism) originates in the atomic and molecular circulation of electrons.

If Faraday was the dominant experimentalist of the nineteenth century, then Maxwell was the dominant theorist. Trying to translate Faraday's pictorial language into mathematical language, Maxwell took all the known laws of electricity and magnetism and attempted to solve them for expressions that would describe electric and magnetic fields. Adding a term to Ampere's law that Ampere had overlooked, Maxwell succeeded beyond all expectation. The measure of his success is that the equations of electricity and magnetism are now known collectively as "Maxwell's equations" though, taken separately, each has its own name; Faraday's law, Ampere's law, Gauss' law and so on. Amazingly, the solutions for electric and magnetic fields in "free space" turned out to be waves. He found that vibrating electric charges produce electric and magnetic fields in the form of waves which travel away from the vibrating source. Even more astonishingly, the speed of the waves was the speed of light! Light, Maxwell concluded, is an *electromagnetic* wave.

Both Faraday and Maxwell firmly believed in the ether. They understood electricity and magnetic forces as forces conveyed by the ether. Light was a disturbance of that medium. However, Maxwell's electromagnetic theory did not explicitly invoke the ether. Hence, the theory did not depend on the existence of the ether.

Published in 1864 and thereafter, Maxwell's theory was not immediately accepted. Not until 1887 was strong supporting evidence for the theory available. It came from the experiments of Heinrich Hertz (1857-1894). Looking for confirmation of electromagnetic waves, Hertz passed a high voltage spark across a wire gap and caused a similar spark to jump across a gap in a metal ring at some distance from the voltage source. Excited by reports of this experiment, the young Guglielmo Marconi (1874-1937) began to dream of transmission of a wireless telegraph, using electromagnetic waves through the ether rather than electric current through a wire. He began experiments in 1894 and made the first transatlantic radio transmission in 1901. Radio waves are also electromagnetic waves that travel at the speed of light.

The accomplishments of Faraday and Maxwell occasioned a major change in physics; the focus on fields rather than particles has been behind many of the advances of twentieth century physics. Einstein's judgment is more than typical; it is definitive.

> The greatest alteration in the axiomatic basis of physics-in our conception of the structure of reality-since the foundation of theoretical physics by Newton, originated in the researches of Faraday and Maxwell on electromagnetic phenomena.[xxii]

Improved measurements of the speed of light were made by several experimenters of the nineteenth century. Arago proposed the first workable terrestrial experiment but went blind before he was able to carry it out. Jean B. L. Foucault (1819-1868) took up Arago's suggestion in 1850 and was able to get an excellent value of the speed of light in air (only 0.6% low). His procedure involved sending a beam of light to a rotating mirror, which, when it moved into proper alignment, sent the beam to a distant, large mirror and it was reflected back to strike the rotating mirror, which had moved in the interim. The return beam therefore struck the rotating mirror at a slightly different spot. From this data, knowing the distance between mirrors, the speed of light was easily calculated.

Foucault's collaborator and friendly rival in this effort was Armand H. L. Fizeau (1819-1896) who devised a wholly different way to measure the speed of light. Fizeau sent light through the gaps in the teeth of a rotating cogwheel to a mirror. When the speed of the wheel was just right, the beam returning from the mirror also passed through a gap in the teeth. From this, Fizeau got good values of the time of flight of the beam.

Knowing the distance to the mirror, he too could easily calculate the speed of light. Fizeau published first, in 1849, Foucault did not actually publish his results until 1862.

Earlier, Foucault and Fizeau collaborated to show that infrared rays are waves. William Herschel (1792-1871) discovered the rays in 1800 by placing a thermometer in a region beyond the red in the rainbow of a prism. The thermometer heated up. Using specially made thermometers, Foucault and Fizeau detected interference of infrared rays.

Foucault also made the first good measurements of the speed of light in water and showed conclusively that it is slower than in air. This result was consistent with light as a wave and was no surprise since the notion of light as a wave rather than a particle was fast becoming scientific dogma.

The American Albert A. Michelson made some simple improvements to the Foucault technique in 1878 and, before his rotating mirror spun off its axis and broke, he obtained a value 0.05% too high. Three years later, he reported two corrections to his results that reduced his error to less than 0.04% too high.

Another Arago suggestion, which he himself was never able to develop, was to detect the presence of the ether by looking for changes in the speed of light from stars as the Earth rotated towards the star and away from it. Attempts at measuring changes in the speed of light traveling through flowing water showed such effects exist. Michelson designed an instrument to detect the tiny variations in the speed of light which the motion of the Earth must produce. The instrument, which at first he called an "interferential refractometer", split a light beam into two beams which traveled over paths of identical lengths at right angles to each other before being recombined. The interference pattern of the combined beams gave an exceedingly sensitive measure of path differences introduced by the medium in which the light traveled. A simple wave of the hand in the air above the instrument produced profound disturbances of the interference pattern. Michelson first attempted the experiment when he was in Potsdam, Germany in 1881. He found no change in the pattern as the instrument was rotated with respect to the Earth and, presumably, the ether around it.

Collaborating with the chemist Edward Morley (1838-1923), Michelson repeated the measurement in 1887. Introducing a number of refinements including floating the entire apparatus in a pool of mercury to isolate it from mechanical shock and vibration (in Potsdam, the stamp of a foot 100 m. from the instrument was detectable) and to allow smooth rotation of the apparatus produced no change in the null result at Potsdam. The Michelson-Morley experiment came to be regarded as the deathknell of the ether. If it had no measurable characteristics, how could it be accorded existence?

To physicists who had finally become quite comfortable with the wave view of light, the Michelson-Morley result was unsettling. There should be a medium of which the waves could be the disturbance. However, the ultimate arbiter of what could be said of light, Maxwell's theory, did not really need the ether. Perhaps all would be well; perhaps the understanding of light would settle down. But that was not to be, more surprises were in store.

The Twentieth Century

The flood of discovery occurring just at the end of the nineteenth century and the beginning of the twentieth profoundly affected our current ideas about light as well as about atoms and matter. The discovery of X-rays, radioactivity and the electron, and studies of blackbody radiation and the photoelectric effect all played a role, as did the invention of the radio and Einstein's relativity.

Special relativity altered our understanding of the relationship between electricity and magnetism. For example, suppose a charged particle moves past us. We see the electric field of the charge but, since a moving charge constitutes a current, we also see the magnetic field of the current.

However, an observer moving with the charge will see only the electric field of the charge because there is no current. One might suppose there is a problem here because different observers see different things. However, this situation is simply one more example of the things that can be altered by the motion of the observer.

One of the first things Einstein did with special relativity was to recast the laws of electricity and magnetism into a form suitable for checking his basic assumptions. The new expression can be easily applied to this example. Of course it gives the right results, but the important consideration at the time was to be sure the laws of physics appear the same to all observers. The reformulated laws of electricity and magnetism passed the test. More than that, they revealed electricity and magnetism as aspects of the same unified behavior. There is a sense in which one might say that a magnetic field is a relativistically transformed electric field.

X-rays, gamma rays, and radio waves all turned out to be electromagnetic waves, differing from visible light only in wave frequency. To this list we can also add ultraviolet and infrared light and microwaves. Physicists call all these by the generic name "light". More is entailed here than a mere extension of lists, however. If these are really all the same thing, why do we have different names for them? The reason is that these different types of *light* interact differently with matter, enabling us to make distinctions and see differences. Trying to understand the differences has led to a deeper understanding of the nature of matter as well as that of light itself.

Planck's study of blackbody radiation was the first sign of a coming revolution. His idea of quantizing the energy of the oscillators emitting the radiation suggested to Einstein that the energy of the emitted radiation itself might also be quantized. In turn, this idea led to his successful explanation of the threshold energy for ejecting electrons from a photoelectric material surface. If the radiation falling on the surface was quantized into packets of exactly the same small quantity of energy, then the energy the electrons received was that amount and no other. Allowing for energy-snatching collisions on the way out of the material implies that there is a maximum in the energy of emitted electrons. Now, the material must hold on to the electrons or they would leak out of it. This energy is called the *work function*. If the energy of the light packets is less than the work function, no electrons can escape.

This explanation describes perfectly what actually occurs. The trouble is that it calls for very unwave-like behavior from light. Waves spread out in space and time. They do not suddenly localize at one spot, give up all their energy and disappear like Einstein's light quanta (photons) do. The localized interaction with electrons and the highly restricted energy make light seem like a particle. But, there is also good evidence showing light behaving exactly like a normal wave. Which is it? Sometimes one and sometimes the other? What kind of reality is that?

Unhappily for those who just wanted light to settle down either as a wave, preferably, or as a particle if it had to, no such resolution is possible. Worse yet, the Davison and Germer experiment on electrons and subsequent similar experiments with neutrons, protons and other subatomic particles makes it clear that this inconvenient, Janus-like two-facedness is not anomalous but is, indeed, a fundamental characteristic of the stuff of matter as well as of light.

The quantum ideas of Planck, Einstein and Bohr led Heisenberg and Schrodinger to create quantum mechanics, the most successful of all physical theories. It accounts for the behavior of electrons and photons, molecules and atoms, solids and even certain fluids. And it accounts for magnetism.

The first step toward a deep understanding of magnetism was taken without any intention of explaining magnetism. Paul A. M. Dirac (1902-1984) was another occupant of the Lucasian Chair of Mathematics at Cambridge and equally as shy as Newton (he ran off to the zoo the day of his appointment to the chair just to avoid being congratulated). In 1927, Dirac became concerned that the Schrodinger equation was not compatible with relativity. A relativistic formulation of the Schrodinger equation, the Klein-Gordon equation, failed to produce a correct spin quantum number for the electron. Spin is the misleading name given the quantized magnetization of a particle. Dirac found a way of rewriting the Schrodinger equation as two equations that gave the correct values of electronic spin. This showed that spin is a relativistic characteristic, and also predicted the existence of the positron. Spin is fundamental to magnetization in materials but is an attribute of electrons and other particles without any relationship to the circulation of electrons around the atomic nucleus. The source of *ferromagnetism* (the permanent magnetism found in iron and other materials) is more deeply hidden than even Fresnel's molecular currents.

In the 1940's and following, a number of people contributed to the quantum mechanical explanation of ferromagnetism. Our present picture is that a quantum mechanical force called the "exchange force" encourages the spins of electrons in neighboring atoms to line up in the same direction. The effect is that of many tiny magnets all aligned so their magnetic fields support and strengthen each other into a strong, externally detectable magnetism. Heat tends to disorder the alignment and reduce the magnetization. The spins in different parts of a piece of iron can also be locally aligned but the local alignment may oppose local alignment elsewhere. This can also reduce the overall magnetization. A sharp blow to a magnet can trigger sudden changes in the local alignments (the domains). Hence, the two old observations that heat and the blow of a hammer can destroy magnetization in iron are now understood.

As to light, it is now being used to test the validity of quantum mechanics and to elucidate the strangeness of quantum mechanical predictions. A long running disagreement between Einstein and Bohr on just how quantum mechanical ideas were to be understood gave rise to the question of whether or not quantum mechanics provides a complete description of things. Bohr thought it did; Einstein thought there were "hidden variables" that, if known, would enable complete, deterministic description of things. In 1975, John Bell (1928-1990) proved that there must be very specific differences in the predictions of quantum mechanics and *any* hidden variable theory about, for example, the polarizations of photons in certain types of experiments. Experiments in the United States and in France have shown conclusively that quantum mechanical predictions are the correct ones. Hidden variable theories are wrong.

While it is comforting to have a well-confirmed theory, a number of the aspects of a quantum mechanical view of the world are not so comfortable. This is especially true in the area of the wave-particle duality of light and of material particles like electrons and protons. It is precisely this area where the new experiments are at once most decisive and most bewildering. They are decisive in the sense that they clearly rule out hidden variable theories and confirm quantum mechanics. They are bewildering, however, when one tries to apply normal physical intuition to them. Quantum mechanical language describes and explains them correctly but translating the mathematical language of quantum mechanics into normal human discourse leads to trouble.

For example, light sources are now available that emit a single photon at a time (as you might expect, they are exceedingly dim). In 1986, following a proposal of Einstein's, Alain Aspect, Philippe Grangier, and G. Roger sent light from a single photon source to a half-silvered mirror. A half-silvered mirror reflects half the photons that strike it and transmits the rest. Thus, half the photons go directly through the mirror to a photographic film beyond and the other half reflect off the mirror. The latter photons then encounter a second mirror that reflects them to the same region of the film as the photons that pass through the half-silvered mirror. After the source has been on long enough to expose the photograph, it is found that a normal interference pattern has developed on the film.

In this single-photon experiment, each photon follows only one path as can be shown by putting detectors on the two paths. When one detector registers the detection of a photon, the other does not; never do the detectors register simultaneously. Also, each photon goes through the apparatus alone. But there is an interference pattern, implying that each photon interferes with *itself!* It looks like the photons are very obliging. *When you set up to detect particles, you detect particles. When you set up to detect waves, waves are what you see.*

This peculiar state of affairs can be pushed further. The decision as to whether to look for a particle or a wave can be delayed until well after the photon has struck the half-silvered mirror. Common sense tells us the photon will have to "make up its mind" at the half-silvered mirror. Either it will be a particle and take one path only or it will be a wave and follow both paths (whatever that can mean). Quantum mechanics predicts, and experiments have confirmed, that the delayed decision does not matter. Particles appear if one looks for particles, wave behavior appears when one is looking for waves.

We thus end our story, having come almost full circle. The sense of awe with which early thinkers beheld light was weakened and then destroyed by mechanistic, particle views of light. Then, the no less mechanistic wave theory superseded the particle view. Now we find ourselves forced to accept a view of light that our language cannot adequately describe. If light is no longer transcendent, it is still ineffable, still a source of wonder.

We are almost back to where we started in another sense, for we are faced with a dilemma the Pythagoreans might well recognize. They began with the intent to see the order of the world in terms of counting numbers and rational numbers. Their efforts led them to discover the irrational numbers, confronting them with a choice, give up the demand for rationals and find a way of coping with irrationals or just give up. We know what course they chose. Similarly, the wave-particle duality confronts us with a choice. Those concepts we want impose on the world, "path", "particle", "wave" and many others, have served us well but we now know they do not always work. What does work is not irrational but is outside our conceptual, linguistic framework. Irrational numbers have now been incorporated into mathematics; they are numbers too. Is there a way to "make sense" of quantum mechanical concepts, a way of seeing that makes the wave-particle duality a part of a seamless whole? There is no answer yet. The story is not ended.

Further Reading

"Andre-Marie Ampere", L. Pearce Williams, *Scientific American*, Jan. 1989, pp. 90-97.

Benjamin Franklin's Science, I. Bernard Cohen, Harvard University Press, 1990.

Catching the Light, The Entwined History of Light and Mind, Arthur Zajonc, Bantam Books, 1993.

Faraday as Natural Philosopher, Joseph Agassi, University of Chicago Press, 1971.

The Master of Light, a biography of Albert A. Michelson, Dorothy M. Livingston, Charles

Scribner's Sons, 1973.

"P.A.M. Dirac and the Beauty of Physics", R.Corby Hovis and Helge Kragh, *Scientific American*, May 1993, pp. 104-109.

"The Science of Optics", David C. Lindberg in *Science in the Middle Ages*, ed. David C. Lindberg, Univ. of Chicago Press, 1978.

About the Author

Dr. John Cramer is Professor of Physics at Oglethorpe University in Atlanta, Georgia. He has almost thirty years of experience teaching undergraduate physics and physical sciences and has authored numerous popular science articles. He is the author of the book How Alien would Aliens be?

Glossary

aether
>
> in Greek Mythology, the clear upper air breathed by the gods on Mt. Olympus. By extension, the matter in the heavens (in geocentric cosmologies) or any sublime or intangible or exquisite matter.

atom
>
> the smallest particle of an element that retains the basic characteristics of the element. The modern view is an atom consisting of a tiny, dense, positively charged nucleus made of protons and neutrons with a number of negatively charges electrons circling outside it. In a neutral atom, the normal atom, the number of electrons equals the number of protons.

calx
>
> the crumbly remains of a roasted (calcined) mineral or metal, an old word for an oxide.

compound
>
> a material composed of identical molecules. Compound is to molecule as element is to atom.

data
>
> (a plural noun, the singular is datum)-facts or measurements, items of sensory experience.

description
>
> an account of an event or events in which data are presented at face value without explanation.

element

> one of the small number of materials from which all other materials are made. An element cannot be reduced chemically to any other materials. Note that an amount of an element is made up of many atoms of that element.

evidence

> information based in sensory experience and capable of being connected to a theory or an explanation of other such information.

explanation

> an account of an event or events in which data are interpreted and fitted together to reveal a coherent pattern or whole.

ion

> an atom with too many or too few electrons to be a normal, neutral atom. Hence, an ion has an electric charge, unlike a neutral atom.

isotope

> (of an element) refers either to an individual atom or to a material made of identical atoms where the number of neutrons in the nucleus is different from the number of neutrons of other atoms of that element.

molecule

> atoms bound together by chemical bonds (bonds involving the electrons of the atoms) into a single item that behaves as a unit. A molecule is the smallest particle of a chemical compound that has the essential characteristics of that compound.

phlogiston

> (from Greek for "inflammable") a hypothetical fluid once thought to be the part of any burning substance that separated out as flame.

prove

give a proof.

proof

a deductive argument leading to a conclusion. The conclusion is then said to be proved. radiation not necessarily connected with radioactivity, this word applies to anything emitted radially outward (in a straight line) from a source.

radioactivity

the emission of radiation by atomic nuclei, specifically, the emission of alpha, beta and gamma rays.

theory

a set of interrelated ideas about certain data or events which explain the data or events. Note that a theory is more elaborate than a single idea or guess. It often involves a model of how events should occur.

Index

Notes

i *Seven Ideas that Shook the Universe*, Nathan Spielberg and Bryon D. Anderson, John Wiley & Sons, 1985, pg. 21.

ii *A History of Western Science*, A.M. Alioto, Prentice-Hall, Inc., 1987, p. 58

iii *De praescriptione haereticorum* chap 7, 9 Quintus Septimus Florens Tertullianus.

iv *A History of Western Science*, A.M. Alioto, Prentice-Hall, Inc., 1987, p. 120.

v *The Sleepwalkers*, Arthur Koestler, Grosset & Dunlap, New York, 1959, p.225.

vi Epigram on Sir Isaac Newton. Alexander Pope,

vii *Forever Undecided*, Raymond Smullyan, Knopf, 1987

viii *De Magnete*, trans. P. Fleury Mottelay, John Wiley and Sons, opening lines of the preface.

ix *Timaeus and Critas*, Plato, trans. Desmond Lee, Penguin Books, 1977.

x *Metaphysics*, Book XII sec. viii). Aristotle.

xi *The Almagest*, trans. R.C. Taliaferro, *Great Books of the Western World*, 1952, vol. 16, p. 270

xii *Ibid.* p. 10)

xiii Isaac Newton in a letter to the Rev. Richard Bentley

xiv Mediaeval Studies IV, ed. A.D. Menut and A.J.Denomy, Toronto: Pontifical Institute of Mediaeval Studies 1941-43, p. 171

xv *Dialogue Concerning the Two Chief World Systems*, Galileo Galilei, trans. Stillman Drake, Univ. of California Press, 1970.

xvi *Galileo Heretic,* Pietro Redonde, trans.Raymond Rosenthal, Princeton Univ. Press, 1987.

xvii *De Rerum Natura*, Book II, Lucretius, Penquin Books, 1962, pp. 63, 64.

xviii "Imitations of Horace" II.i., Alexander Pope.

xix "The Second law of Thermodynamics" by Rudolf J. E. Clausius in *A Source Book in Physics*, W.F. Magie, Harvard Univ. Press, 1963, p. 234.

xx the *Timaeus*, Plato, trans. Desmond Lee, Penguin Classics, pp. 62-63)

xxi *The Nature of the Universe*, Lucretius, trans. Ronald Latham, Penguin Classics, Book IV),

xxii ("Maxwell's Influence on the Development of the Conception of Physical Reality", by Albert Einstein in *James Clerk Maxwell*, J.J. Thomson, Macmillan, 1931, pp. 66-67).

www.ingramcontent.com/pod-product-compliance
Lightning Source LLC
Chambersburg PA
CBHW020742180526
45163CB00001B/312